David Prince

Orthopedics

A systematic treatise upon the prevention and correction of deformities

David Prince

Orthopedics
A systematic treatise upon the prevention and correction of deformities

ISBN/EAN: 9783337085216

Printed in Europe, USA, Canada, Australia, Japan

Cover: Foto ©berggeist007 / pixelio.de

More available books at **www.hansebooks.com**

ORTHOPEDICS:

A SYSTEMATIC TREATISE

UPON THE

PREVENTION AND CORRECTION

OF

DEFORMITIES.

"In the practice of Surgery, we have more frequently to lament the inefficacy of the art, from the advanced stage in which diseases are presented to us, than from anything incurable in their nature."—THOMAS COPELAND.

BY

DAVID PRINCE, M.D.

PHILADELPHIA:

LINDSAY & BLAKISTON.

1866.

SHERMAN & CO., PRINTERS

PREFACE.

THIS treatise has been prepared with special reference to the wants of physicians engaged in general practice.

Outside of the large cities, the majority of patients needing attention on account of deformities, or diseases or accidents that lead to them, must find relief at the hands of the profession in their near vicinity, or not find it at all.

The proportion of the wealthy, who can be sent a considerable distance to specialists, is so small in relation to the whole, as to make very little diminution of the number that must be treated at home.

This renders it important that there should be a general prevalence, in the profession, of a knowledge of prevention and treatment; to bring the means of relief within the pecuniary resources of the majority of sufferers.

In order to render the advance in knowledge on this subject, gained within the last twenty years, accessible to the mass of the profession, it is necessary that the substance of many valuable essays and monographs should be sifted and collected into a few pages, within the means of all to purchase it; and within the time of all to read it.

To these ends, an attempt has been made to connect the medical treatment with the mechanical, in order to give the work its nearest practical approach to completeness, compatible with the necessary brevity.

As far as possible, proper credit has been given to the authors of recent advances, both in the science of the subject, and in the invention of expedients to meet indications.

Those who have time for research, will find many subjects treated much more in detail, in the contributions referred to, as the sources from which much of the materials of these pages has been derived.

JACKSONVILLE, ILLINOIS, Sept., 1866.

TABLE OF CONTENTS.

	PAGE
DEFINITION, ...	17

PART FIRST.

CLASSIFICATION, ...	19
I. Arrest, Redundancy, or Misplacements of Development, .	20
1. Arrest, ...	20
Hare-lip—Cleft Palate—Obturators—Artificial Palates —Hypospadias—Defective Closure at Umbilicus, &c.,	24
2. Redundancy, ...	26
3. Misplacements, ...	26
II. Perversion of Relations of Parts through Muscular Contraction, generally owing to Palsy of Nerves distributed to one set of Muscles, or to Irritation of those distributed to their Antagonists, ...	26
Nervous Irritation, ...	27
Contracture, ...	28
Muscular Rigidity, ...	29
R. B. Todd's Views, ...	29
Brown-Sequard's Views, ...	31
Treatment, ...	34
Inferences drawn from Experiments with Galvanism, .	35
Effect of Over-distension of Muscles, ...	35
Effect of Division of Muscles and their Tendons, . .	37
Passive Motion, ...	39
Experiments of Reid, ...	40
Brown-Sequard's Theory of the Action of Electricity, .	42

PAGE

MECHANICAL TREATMENT OF PALSY AND PERVERSION BY
   IRRITATION, ... 43
   1. To Maintain the Normal Relations, ... 43
   2. To Protect Partially Paralyzed Muscles from Over-
      action (Writers' Cramp), ... 44
   3. To Keep Up the Circulation, ... 45
STRABISMUS, ...
   Operation, ... 46
   Exercise of Each Eye Separately, ... 47
   Stereoscope, ... 48
   Plane Prisms, ... 48

III. Inflammation, and Perversion of Nutrition, Injuring or De-
   stroying the Tissues, Perverting or Nullifying Muscular
   Contraction, Softening Bone, Distorting or Stiffening Joints,
   and Perverting, Impairing or Abolishing the Functional
   Movements of the Parts Affected, ... 49
   Synovitis, Varieties, ... 50
      "   Symptoms, ... 51
      "   Pathology, ... 52
   Pressure the Cause of Exfoliation, ... 54
   Treatment, ... 58
   Principles of Abernethy, ... 59
   Counter-irritation, ... 60
   Suppuration—Question of Opening, ... 61
   Extension, ... 62
   History of the Remedy, ... 62
   Principles Governing its Employment, ... 66
   Distortions and Stiffenings after the Subsidence of In-
      flammation, ... 67
   Difficulties to be Overcome, ... 68
   Extension to be Practised Before Attempting to Change
      the Angle, ... 69
   Division of Muscles and Tendons, ... 72
   Extreme Extension Practised Under Anæsthesia, . . 73
   Barton's Operation, ... 73

IV. Accidents, Breaking, Tearing, Bruising or Freezing the Tissues, followed by the Loss of the Parts, or the Failure, Imperfection or Perversion of their Union or Restoration, . 74
   Prevention of Deformity More Easy than Removal, . 74
   Softening Bones by Drilling, ... 74

PAGE

Cicatrices Difficult to be Elongated, ... 75
Rickets, &c., ... 77

V. Mutilation of Parts Designedly Done, or Force or Restraint Artificially Applied, like the Compression of the Feet of the Female Children of Chinese Grandees, the Heads of some American Indian Tribes, and to a Less Degree the Feet and Waists of Genteelly Educated Children in Modern American and European Society, ... 77

PART SECOND.

PARTICULAR DISEASES AND DEFORMITIES NOT YET NOTICED, OR
ONLY INCIDENTALLY REFERRED TO, ... 79
Hip Disease, ... 79
Diagnosis, ... 79
Dislocation, ... 81
Apparent Dislocation, ... 82
Anchylosis, ... 83
Inflammation of the Knee-joint, ... 84
Inflammation of Other Joints, ... 85
History of Modern Apparatus, ... 86
Attachment for Extension, ... 94
Lateral Curvature of the Spine, ... 96
Classification, ... 97
1. Weakness of Bones, ... 98
2. Diseases of Ligaments, ... 99
3. Weakness of Muscles, ... 100
4. Spasmodic Contraction, ... 104
Wry Neck, ...105
Treatment, ... 106
Mode of Division of Sterno-Mastoid Muscle, . . 107
Apparatus, ... 108
Treatment of Confirmed Lateral Curvature, . . . 109
Exercise Alone Insufficient, ... 110
Elasticity in Appliances, ... 111
Mechanism, ... 112
Antero-posterior, Vertical or Angular Curvature—Pott's
Disease—Kyphosis, ... 126
Pathology, ... 127
Symptoms, ... 128

PAGE

Historic Development, ... 129
Theory of the Deformity, ... 133
Treatment, General, ... 134
 "   Local, ... 135
Purulent Accumulation, ... 136
Quietude, ... 138
Artificial Support, ... 139
Shelldrake's Apparatus of 1782, ... 140
Modern Application of the Views of Shelldrake and
  Earle, ... 141
Time, Exercise and Food, ... 148
Atmosphere, ... 150
Talipes, ... 151
Definition and Classification, ... 151
Talipes, Equinus, ... 151
 "   Dorsalis, ... 153
 "   Varus, ... 154
 "   Valgus, ... 156
 "   Plantaris, ... 157
 "   Calcaneus, ... 159
Complications, ... 160
Enumeration of Causes of Talipes and Allied De-
  formities, ... 160
Spasm and Palsy, in their Relation to Talipes, . 165
Treatment, ... 169
The Hand the Type of Apparatus, . . . 170
Operations, ... 172
The Results of Tenotomy—Adams's Statistics, 176
Paget's Observations, ... 177
Apparatus and Tenotomy, ... 178
Mechanical Treatment, ... 187
Plan of Hippocrates, ... 188
Modern Plans, ... 189
Barwell's Method of Adhesive Plaster Attach-
  ment to the Leg, ... 193
Difficulties to be Overcome in T. varus, . . 195
Gutta-percha Attachment to the Foot, . . 200
Iron Shoes Thrown Away, ... 206
Brace for Weak Ankle, ... 207
Deviations at the Knee-Joint, ... 208
Knock-Knees and Bow-Legs, ... 209
Upper Extremity, ... 210
Traction upon Thumbs and Fingers, . . . 210

	PAGE
Difficulties Attending Division of Tendons,	210
Deformities Requiring Plastic Operations, Cicatrices,	211
A Line of Immobility may be Produced,	212
Deformities After Fractures, Avoidable and Unavoidable,	213
May Occur Subsequently to Union,	213
Prevention,	213
Correction,	214
Enumeration of Measures,	215
Delay or Failure of Union,	216
Enumeration of Measures,	218
Drilling,	219
Malgaigne's Spike,	220
Cases,	223
Drilling the Callus Alone Unsuccessful,	227

B

LIST OF ILLUSTRATIONS.

FIGS.   PAGE

1, 2, 3, 4, 5. Artificial palate (Kingsley),   .   .   .   22, 23
6. Galvanic excitation, from Brown-Sequard,   .   .   .   36
7. Support for weak lower limb (Bigg), ... 44
8. Plan for exercising a palsied limb (Bonnet), ... 45
9. Exfoliation of hip-joint (Andrews), ... 55
10.   "   of knee-joint (Andrews), ... 56
11. Dislocation of knee-joint (Tamplin), ... 68
12. Apparatus for reducing flexure at the hip-joint (Andrews),   .   69
13.   "   for the knee-joint (Andrews), ...   70
14.   "   for rotating the forearm (Bonnet), ...   71
15. Apparent dislocation at the hip-joint, ... 83
16, 17, 18, 19, 20, 21, 22, 23, 24. Davis's splint, .   .   . 86 to 91
25. Sayre's splint, ... 91
26. Barwell's splint, ... 91
27. Andrews's splint, ... 92
28. Prince's splint, ... 93
29. Barwell's adhesive plaster attachment, ... 95
30. Lateral curvature of the spine (Bonnet), ... 97
31. Torticollis (Bonnet), ... 104
32. Apparatus for Torticollis, ... 107
33, 34. Apparatus (Bonnet), ...   108, 109
35. Stretch-bed (Bigg), ... 112
36. Delpech's extension, ... 113
37. Spinal splint (Andrews), ... 114
38.   "   (Tamplin), ... 116
39.   "   (Bigg), ... 117
40. Tavernier's belt, ... 117
41. Lateral support (Andrews), ... 118
42.   "   (Davis), ... 118
43, 44. Spinal chair (Bonnet), ...   119, 120
45, 46.   "   (Andrews), ...   121, 122
47, 48. Extemporized Apparatus, ... 123
49. Lateral curvature, ... 125
50, 51. Vertical curvature (Cruveilhier), ...   131, 132

FIGS.   PAGE
52. Shelldrake's apparatus of 1782, ... 140
53. Davis's spinal splint, ... 142
54, 55. Andrews's " 142, 143
56, 57. Bonnet's spinal shield, ... 144
58.   " for suspension, ... 145
59. Extemporized apparatus, ... 146
60. Extension of Ambrose Paré, ... 148
61. Talipes equinus, ... 152
62.   " dorsalis, ... 153
63.   " equino-dorsalis, ... 153
64, 65. " varus, ... 154
66, 67. " valgus, ... 157
68, 69, 70. Talipes plantaris, ... 158
71. Talipes valgo equinus, ... 158
72.   " calcaneus, ... 159
73. Illustration from Little, ... 165
74. Scarpa's iron frame, ... 190
75. Scoutetten's apparatus, ... 190
76. Dr. Post's plan of applying gutta percha, ... 191
77. Andrews's dressing for T. varus, ... 193
78, 79, 80. Barwell's method for Talipes, ... 197, 198
81, 82. Gutta percha attachment to the foot, ... 203
83, 84. Before and after treatment, ... 204
85. Brace for deviation at the ankle, ... 207
86. Apparatus for knock-knees, ... 209
87.   " for bow-legs, ... 209
88. Modification of Malgaigne's spike, ... 222
89, 90. Deformity after fracture, ... 225
91. Apparatus for reducing angularity of bone, ... 226
92. Deformity with nonunion, ... 228
93. Same, restored, ... 228

ORTHOPEDICS.

DEFINITION.—*Orthopedics*, **Orthopædia**, *Orthopedic Surgery* (from Gr. *orthos*, right, and *paidos*, genitive for child). The part of medicine whose object is to prevent and correct deformities in children. Used with a more extensive signification, to embrace the prevention and correction of deformities at all ages.*

Prevention, in this as in other branches of medicine, is far more important than correction, though usually less appreciated, because the extent of deformity obviated can only be estimated by those familiar with the subject. The impossibility of correcting many deformities, which, by timely care, might be prevented, adds still more to the necessity for a full account of the treatment of those diseases and injuries from which deformities result. The necessary limit of this treatise has made a very full account of these lesions impracticable, but enough is said to indicate their character, and the sources from which fuller information may be derived.

The efficacy of time, skill, and patience, in correcting malformations, is not only far beyond popular credence, but beyond the estimate of the majority of the medical profession, who have not acquainted themselves with the recent advances in the art. The length of time often required is sometimes a

---

* The term was introduced by M. Andry, who was Dean of the Medical Faculty of Paris, and who, in 1741, published a work in two volumes on "Orthopedie, or the Art of Preventing and Curing Deformities in the Human Body."

2

source of discouragement with those who do not reflect by what slow and insidious processes deformities are often produced. The popular idea that surgery is cutting and curing quickly, and the impatience of surgeons themselves who share this estimate of the art, are impediments in the way of the accomplishment of that degree of success which the general practitioners in every community should be able to secure.

THE following classification may aid in securing a clearer conception of the modes in which malformations or malpostures originate, and of the means appropriate to their prevention and correction.

CLASSIFICATION OF DEFORMITIES BY THEIR CAUSES.

I. ARREST, *redundancy, or misplacement of development. Congenital.*

II. PERVERSION *of relations of parts through muscular contraction, generally owing to palsy of the nerves distributed to one set of muscles, or to irritation of those distributed to their antagonists. Con or post-genital.*

III. INFLAMMATION *and perversions of nutrition, injuring or destroying the tissues, perverting or nullifying muscular contraction, softening bones, distorting and stiffening joints, and perverting, impairing, or abolishing the functional movements of the parts affected. Generally post-genital.*

IV. ACCIDENTS, *breaking, tearing, bruising, burning, or freezing the tissues, followed by the loss of the parts, or the failure, imperfection, or perversion of their union or restoration.*

V. MUTILATION *of parts designedly done; or force or restraint artificially applied, like the compression of the feet of the female children of Chinese grandees, or the heads of some Indian tribes, and, to a less degree, the feet and waists of genteelly educated children in modern American and European society.*

I. Arrest, Redundancy, or Misplacement of Development.

It is impracticable here to enter into a full consideration of the varieties of congenital deformities: only a few can be mentioned which admit of remedy.

1. Arrest.—The most common instance of arrest is that of harelip, which is an arrest of the process of union of the integument and other components of the upper lip to one side of the median line corresponding with the junction of the os maxillare with the os incisivum, which in man become one bone, though they remain distinct in quadrupeds. The extension of the fissure between these two bones, and along the median line between the maxillary and palate bones, makes a fissure of the hard palate, and the still further extension through the soft palate, produces a cleft palate. The latter may exist alone while the lip and hard palate have their proper development. The union of opposite surfaces in the fissure of the lip is easily accomplished, that of the palate with more difficulty, while the union of the separated parts of the bony arch of the mouth has resisted the attempts of surgery, owing to the firmness of the bones of the face.

Some diminution of the breadth of the fissure may have been secured by pressure upon the malar bones in early infancy, and Dr. J. Mason Warren, of Boston, has done something toward the closure of this opening by dissecting the soft tissues with more or less of the periosteum from the bony arch, and causing them to glide backward and inward to fill the fissure to a greater or less extent.

The success in closing the fissure in the soft palate has been very creditable to the surgical art, but the restoration of the function of the palate has been so seldom and so imperfectly achieved as to leave the utility of the operation very much in doubt.

One reason of failure is probably the late period at which the operation of staphyloraphy is usually and almost necessarily performed, the parts having lost their instinctive or automatic aptitude, and another is the tension of the curtain of the palate from deficiency in the amount of the material, some of which, scanty as may be the supply, is necessarily lost in the preparation of the edges of the fissure for adhesion.

From one or both of these causes, it becomes, in most cases, impossible to bring the velum in contact with the posterior wall of the pharynx in order to shut off the expired air from flowing freely out through the nostrils. The execution of many of the elements of articulate language thus remains impossible.

In view of these imperfections incident to the results of closure of the soft palate by surgical operation, De Le Barre, about the year 1820, constructed an artificial palate of caoutchouc, which he said answered the purpose of articulation.

Long ago, even during the period of Grecian civilization, wax and other ingredients had been used to close the communication between the mouth and nose; and in the sixteenth century, Ambrose Paré invented an appliance which he described as " A gold or silver plate, larger than the aperture, of an arched form, and middling thickness, having attached to the surface that answers to the nose, two stems a few lines long and a piece of sponge about the size of the orifice placed between the two stems. The instrument being fixed, the moisture of the part makes the sponge swell, which then becomes locked in the aperture and maintains the plate in position." Heister also described a sponge obturator, but De Le Barre was the first to construct one of elastic rubber.

But little attention seems, however, to have been paid to De Le Barre's improvement, and, about 1845, Dr. C. F. Stearns, an American surgeon, made for his own use an instrument of the improved soft elastic vulcanized rubber, which produced such results as at first to seem all that was desired, but so complicated was it in the number and connection of its parts that it was very liable to get out of order, and it was probably never successfully and permanently worn by any one but himself.

Taking advantage of the good points in the contrivance of Dr. Stearns, rejecting its defects and complications, and replacing them with simple expedients not liable to derangement, Dr. Norman W. Kingsley, Professor of Dental Mechanism in the New York College of Dentistry, has more recently constructed an elastic rubber palate which is rapidly coming into general favor. The artificial palate, with its projections to fit the form of the parts, is in one piece of soft rubber, retained from sliding back by a single gold or hard rubber plate, attaching to one or more front teeth; the whole being so arranged as to be easily removed for cleanliness or other purposes, and replaced by the wearer.

The general form of the ingenious appliance is explained in connection with the figures.

The rubber is vulcanized in a mould made accurately to represent the shape of the parts, so that though a single palate may last only a few months, duplicates can be made at very trifling cost.

The closure of the fissure in the roof of the mouth (when it exists) by elastic material and the supplying of a vibrating soft palate in place of the natural one which is absent, not only adds to the comfort of the patient, improving his facility of swallowing, but it improves the phonation immediately and

Fig.1

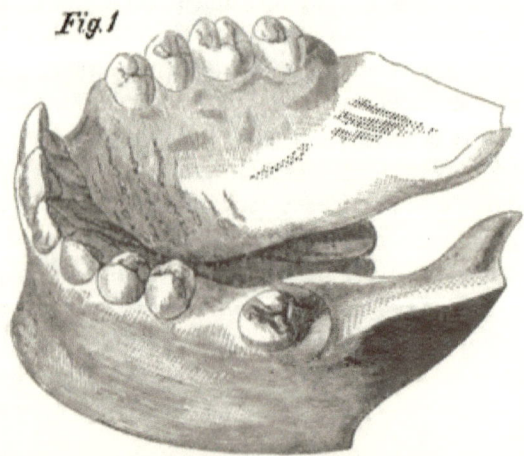

Fig. 1 represents a model of a fissured palate, complicated with hare-lip on the left of the mesial line. There is a division also of the maxilla and the alveolar process: the sides being covered with mucous membrane which come in contact with each other, but are not united. The left lateral incisor and left canine tooth are not developed.

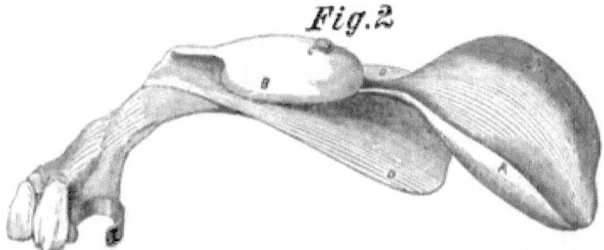

Fig. 2

Fig. 2 represents the artificial velum, as viewed from its superior surface, together with its attachment and two artificial teeth to fill the vacancy.

The lettered portion of this appliance is made of elastic vulcanized rubber; its attachment to the teeth of hard vulcanized rubber, to which the velum is connected by a stout gold pin, firmly imbedded at one end in the hard rubber plate. The other end has a head, marked C, which being considerably larger than the pin, and also the corresponding hole in the velum, it it forced through—the elasticity of the velum permitting—and the two are securely connected.

The process B, laps over the superior surface of the maxilla (the floor of the naris), and effectually prevents all inclination to droop.

The wings, A A, reach across the pharynx, at the base of the chamber of the pharynx, behind the remnant of the natural velum.

The wings, D D, rest upon the opposite or anterior surface of the soft palate.

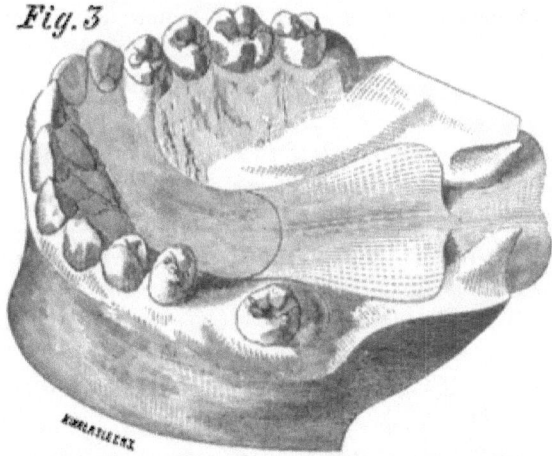

Fig. 3

Fig. 3 represents a model the same as Fig. 1, with the appliance, Fig. 2, *in situ.*

The wings, D D, in Fig. 2, and the posterior end of the artificial velum only, in this cut being visible.

Fig. 4.   Fig. 5.

Transverse section symmetrical.   Transverse section unsymmetrical.

the articulation by practice.   The effect of the appliance on phonation and articulation was very satisfactorily demonstrated by Dr. Kingsley, by trained subjects of cleft palate, before the American Medical Association at its meeting in New York in 1864, and before the New York Academy of Medicine in October, 1865.

Without claiming perfection for the substitute for the palate, Dr. Kingsley has found " a decided improvement in speech within a very few weeks, a clearness and distinctness of utterance which the patient never showed before.   In some cases this progress has been so rapid that, within a period of six months after its first introduction, the wearer would not from his speech be suspected by the ordinary observer of possessing such a deformity; and in every case, within a few months, the speech has improved so much as to render it perfectly intelligible to strangers without repetition.   It must not be supposed from these statements that there is any marked change in articulation immediately.   There is almost always an immediate change in the *tone of the voice*, which is much less disagreeable than formerly, and this change is often mistaken for an improvement in articulation."*

HYPOSPADIAS, an open state of the urethra from arrest of development, admits of some improvement by plastic operations securing artificial union; but here, and elsewhere, the defects must be supplied out of material sufficiently like the original to fulfil its function.

Defective closure at the umbilicus and in the inguinal canals leading to hernia, is usually closed subsequently by the pressure and irritation of appropriate trusses perseveringly worn.   The principle of treatment is to prevent the exit of the intestines or omentum, and thus to favor the contraction of the tissues around the opening, and by pressure to secure final adhesion

* See Observations on Obturateurs or Artificial Palates, by James Snell, Surgeon-Dentist; London, 1824.   London Lancet, 1845.   Transactions of the London Odontological Society, 1856–57.   Edwin Searcomb in Medico-Chirurgical Transactions; London, 1857.   Transactions American Medical Association, 1864.   Richardson's Mechanical Dentistry, p. 392, from Vulcanite for 1860, and Bulletin N. Y. Academy of Medicine for January, 1866.

of adjoining peritoneal surfaces. The radical cure of inguinal hernia has been attempted by approximating the pillars, as in the operation of **John Wood**; and by filling the neck of the sac, by invaginating the superficial tissues, as in the original operation of Wutzer, and its many modifications.

A few years ago a man visited various parts of the United States exhibiting to medical men a deficiency of the anterior wall of the bladder, and the corresponding portion of the abdominal wall, so that the posterior wall of the bladder occupied the place of integument, and the ureters opened directly upon the surface. There was at the same time, only a diminutive and imperforate penis. This man was in all other respects well formed.

Mr. Bigg, in his " Orthopraxy," describes a similar case, for which he constructed a receptacle for the urine, which greatly increased the comfort of the patient.

Deficiency of a portion of the posterior bony wall of the spinal canal (spina bifida) giving rise to a hernia containing the cerebro-spinal fluid, is a much more common deficiency from arrest of development.

From the generally fatal result under all methods of treatment, as well as when the case is left to nature, this is truly an opprobrium in surgery. Where the neck or pedicle of the sac is narrow, the adhesion of its opposed surfaces may be sometimes secured by injection of iodine or other irritating liquid, or by scarifications managed so as not to admit air into the wound, and so as to secure union of the cutaneous incision by primary adhesion, and by ligature; but the frequent occurrence of a rapidly fatal inflammation of the arachnoid membrane has led to the general preference for carefully applied continuous pressure.

Another variety of arrest consists in the occurrence of deep fissures in the limbs and even their complete amputation, apparently occurring at a late period of intra-uterine development, and depending upon mechanical restraint external to the fœtus.

There is the greatest probability that solidifying exudations from the interior of the amniotic membrane may, for a variable period, take the **place of** the ordinary fluid secretion. Bands formed of this material may, in some cases, produce amputation ; in others, **fissures, in** some instances dividing muscles, leaving the parts to be united by tendinous connections. For better appearance, fissures may often be closed by converting them into the conditions of incised wounds, approximating the surfaces, retaining them in apposition for adhesion.

2. REDUNDANCY, as in supernumerary **fingers and toes,** usually admits **of an easy** remedy in amputation, which should be so practised as to leave the parts as nearly in their natural condition **as** possible; a point requiring no little surgical skill. **Redundancy,** in the uniting process between adjoining parts, **gives rise to web** fingers **and** toes. With the latter, incased **in** a shoe, the deformity hardly amounts to an inconvenience, but in the **fingers the** abnormal union is both inconvenient and unsightly. **The fingers may be** dissected apart, but the deficiency of integument is apt to lead to a reunion by a gradual encroachment of the fork of the fingers, very much like the process after a burn of the same parts. The sacrifice of the bones of a finger for the use of its integument to apply to the adjoining fingers, renders the operation more promising.

3. MISPLACEMENTS not only constitute a very numerous variety of congenital deformities, but the most of them admit of no amendment. One limb **is** grown upon or into another, making **one** member, with all imaginable varieties, sometimes presenting the whole ten toes and fingers; there are sometimes two heads with one body, **or** two bodies, **with** their limbs, sustaining one head, or two distinct beings are united by more or less extensive connections. These rarely admit of any remedy.

II. PERVERSION OF RELATIONS OF PARTS THROUGH MUSCULAR CONTRACTION, GENERALLY OWING TO PALSY OF THE NERVES DISTRIBUTED TO ONE SET OF MUSCLES, OR TO IRRITATION OF THOSE DISTRIBUTED TO THEIR ANTAGONISTS.

The admission of the existence of the **same** kind of deficiencies, or perversions of the **nervous function in** intra- and extra-uterine life, throws much light upon many congenital deformities. While talipes probably often owes its existence to some constraint, analogy leads us oftener to entertain the theory of irregular muscular action, through disease of the nervous system. The frequency of hydrocephalus during intra-uterine growth, shows that the fœtus is liable to the same perversions of nutrition which in childhood produce the same pathological results. For illustration, a contraction of the little finger most, and of the other fingers less, while the fontanelle is seen to be depressed, is sometimes observed as a symptom of the incipiency of those cerebral lesions in insufficiently nourished children, which, when unchecked, end in convulsions and death. A prolongation of this incipient state, previous to birth, or subsequent to it, might easily be supposed to produce such a preponderance of the flexors over the extensors as to require artificial aid to secure restoration.

A change of the state of assimilation by proper alvine evacuants, and an increase or alteration of the nutriment may, in children, remove the cerebral irritation, and relieve the local tonic spasm before the affected muscles become permanently shortened. On the other hand, an inflammatory excitement may have an equally abortive, protracted, or fatal course, with equally varied effects upon muscular contraction. Sometimes these lesions of the nervous system are so local as not apparently to pervert the general health, and yet the proper general treatment often removes the irritation and corrects the abnormal muscular contraction. In children, the progress of these cases can be observed, but in the fœtus the nature of the lesion can be inferred only by the result.

The interest and importance of this subject justify its consideration to a greater extent. Its correct understanding is necessary to the adoption of such therapeutics as will *prevent* deformity or loss of functions, which is far better than the greatest success in restoring them. Any discussion on the subject of orthopedics which should omit this, would be defective in one of the most important points. A correct theory of the nervous irritation has two practical ends,—first, to subdue or ameliorate the original nervous disorder; and secondly, to preserve the functions of the parts receiving the perverted nervous stimulus, so that when the original disease passes by, the local parts shall not have become irredeemably useless by loss of nutrition, or the assumption of malpostures, or both combined.

Great light is thrown upon this subject by the hypothesis, pretty well sustained, that there may be a constant muscular pull in addition to the ordinary muscular tonicity, and independent of, and sometimes in spite of, and in antagonism to voluntary contraction and ordinary reflex action. This constant abnormal contraction results in shortening, rigidity, and ultimate wasting of the muscles affected, reducing them to the condition and function of ligaments—in the language of Barwell, the condition of *contracture*.

This is something more difficult to control than mere palsy or loss of action of certain muscles, leaving their antagonists in the latter case to distort the parts affected by a mere excess of tonic and voluntary power. One of the best and most recent surgical writers in the English language has, however, thrown doubt over this hypothesis, or been disposed greatly to narrow its field. We quote from Barwell on Club-Foot, p. 19. (Churchill, London, 1863.) "Infants are frequently, as is well known, subject to convulsions, and it is averred that sometimes one or more [muscles] which have, during the attack, drawn the limb into malposture, do not recover from the contraction, but continue to keep the limb distorted. Such a muscular condition would not be a *contracture*, as I have described it, since the convulsion never lasts sufficiently long for any such organic process of areolar shortening; the state would be one of persistent, unvarying spasm, powerful enough to overcome the antagonistic healthy muscles, and permanent enough to produce lasting change of form. Such condition does not only never come under our notice, but is, I believe, pathologically impossible. There are, no doubt, a few cases of paralysis of voluntary power over the muscles, while the excito-motory function continues, and in the spasm of the whole set, the strongest will, of course, predominate. **Voluntary** power is as much used to control as **to** excite. The paralysis **of this** power **is** evidenced as much by violent and uncontrollable spasm, **as by** insusceptibility of subordinate movement."

The hypothesis in question **is in** perfect accordance **with the** last part of this quotation—maintaining that in addition to the excito-motory function, there is another which **does not tire** during the continuance **of** the irritation, but constrains **the** muscles affected to draw continually, while the capillaries **are** so squeezed as to receive **an** inadequate supply of blood, resulting in rapid **wasting of the muscular** substance, **from want** of nutrition. **In illustration of this** subject, **the** following **quotations are from Dr. R. B.** Todd's **account** of paralysis with rigidity. **(Todd on the Nervous System,** Lindsay **& Blakiston's edition, 1855. Page 153.)**

After speaking of paralysis, with relaxed muscles, **he says:** "**I mentioned that** the second kind of hemiplegia is **where the** muscles of the paralyzed limbs are rigid, and where the rigidity comes on simultaneously with the paralysis, or very soon after it. You **must bear** in mind that a distinctive feature of **this** hemiplegia **is** the early period at which the muscles assume **this** rigid condition. This is the more important, inasmuch **as we** meet with another form of hemiplegia with rigid muscles, **in** which **the** stiffness gradually supervenes a long time after **the paralytic seizure,** and **may succeed to the relaxed condition of the paralyzed muscles."**

Page 155. "**The condition of slight and partial rigidity of** paralyzed **muscles is that of most frequent occurrence in hemi-** plegia **caused by an apoplectic clot. My idea as to its cause, is that it depends upon a state of irritation, propagated from torn brain to the point of** implantation **of the nerves of the** affected muscles. **But** you will ask, Why is it that, in some cases of clot, the hemiplegia will be accompanied with complete relaxation **of muscles, while in** other cases the rigidity of which I have **spoken exists? The answer to** this question **is as follows: In the cases where** there is no rigidity, the clot lies in the midst of softened brain, **and has not in any degree**

encroached upon sound brain, but when rigidity exists, the clot has extended beyond the bounds of the white softening, and has torn up, to a greater or less extent, sound brain."

Page **158.** "A much more interesting form of the second kind of hemiplegia, is that in which there is considerable rigidity of all the muscles of the arm and forearm, where the arm is kept at an angle with the trunk (and sometimes these patients hold it across the chest), the forearm being flexed on the arm, and the fingers flexed on the palm. In these cases the paralyzed muscles seem to be firm and contracted, and sometimes almost in a tetanic state, and offer considerable resistance to flexion or extension, which frequently also excites a good deal of pain. When the rigidity is of this nature, the paralysis is generally not complete, a certain degree of power remaining of moving the whole limb, or some part of it, and very frequently sensibility is affected, being sometimes obtuse, but oftener in an excited condition, while it not uncommonly happens that reflex actions are also considerably exalted. In the latter cases, you will find the reflex actions produce considerable pain."

Page 159. "When hemiplegia, with a rigid state of the muscles, supervenes soon, on some injury of the head, you may almost always make a certain diagnosis that it is the result of irritant compression of the opposite hemisphere, by depressed bone or hemorrhage, outside or inside of the dura mater."

Page 160. "It (paralysis with rigidity) is due to a cause which exercises at once, a paralyzing and irritating influence on the brain, and this influence is propagated to the spinal cord, and through the nerves implanted in that portion of the nervous centre, to the muscles of the paralyzed limbs, in which it excites a state of contraction. The effect is analogous to that produced by the continued action of the electro-galvanic machine, and just as a rapid succession of electric shocks may keep up this rigid condition of muscles, so may continuous shocks of nervous force, due to irritative pressure, bring about a similar result."

Page 178. "I think it must be admitted that the rigid state

of the muscles is due, primarily, to an irritated or excited state of the nerves, and, that on the cessation of that irritation, the muscles might resume their relaxed condition, or that a similar result would follow a severance of all connection between the muscles and the seat of the cerebral lesion, by section of the nerves. It seems to me, however, that after a long continuance of this rigid and shortened state, the muscles would become permanently shortened, and would assume a condition similar to that into which anchylosed joints are apt to fall — a condition from which they would recover very slowly or not at all."

These detached quotations necessarily appear dogmatic, and for the facts and reasons through which the conclusions are arrived at, the reader is referred to the book from which they are taken.

Brown-Sequard goes a step farther, and claims that this permanent contraction may occur, not only from irritation at the origins of the nerves in the encephalon, but that it may occur from irritation of the appropriate nerves in their course from centre to circumference. He is of opinion that there are nerve-fibres going from the brain to the muscles, the irritation of which produces tonic or permanent contraction of the muscles to which these nerves go, and that these nerve-fibres pass down in portions of the cord different from those through which the voluntary nerve-fibres pass. A few quotations are here introduced from the valuable work of this author, — The Physiology and Pathology of the Nervous Centres, p. 194.

"The division of these nerve-fibres is not followed by paralysis, although they are able to act on muscles to produce contractions, and even more powerful than those caused by nerve-fibres employed by the will in voluntary movements."

"It is a fact, worthy of attention, that a puncture with a needle through the anterior pyramids, which contain very nearly all, if not all, the nerve-fibres employed in voluntary movements, will hardly produce a momentary contraction in some muscles, while certain punctures through the olivary column of the medulla oblongata at once produce a spasm of many muscles, although this column does not contain more than very few voluntary motor fibres, if any at all; and now, to add to the strangeness **of the** fact in this last case, the muscles remain contracted for hours, sometimes for days and **weeks.**"

" **We** have all been taught that after the removal of a cause **of excitation** in the nervous centres, as well as in the nerves, the effects of the excitation disappear, until inflammation supervenes and produces a permanent excitation; while here, however, **we see a** puncture with a needle, or a section with a knife, before any inflammation can have begun, followed by **a** persistent effect. There is, therefore, in some parts of the nervous centres, a property of acting in a persistent manner to produce muscular spasms during, and after a mechanical excitation."

Page 196. " The parts **of** the base of the encephalon which are capable **of producing** persistent **spasms,** seem to be quite different from those employed in the transmission **of** sensitive **impressions, or** of **the orders of the will to the** muscles, at least in the medulla oblongata and póns Varolii. They constitute a very large portion of these two organs, and perhaps three-fourths of the first one. They are placed chiefly in the lateral and posterior columns of these organs; many of their fibres do not decussate and produce spasms on the corresponding side of the body; they seem to contain most of the **vaso-** motor nerves, by which, directly or through a reflex action, they **may act on other parts of** the **nervous system.** They have much to do with the phenomena of **several, if not most** of the convulsive diseases. Lastly, I will say that the history of their properties **and actions** throws a great deal of light upon the effects of disease or extirpation of the cerebellum."

Page 207. " There are a great many nerve-fibres and **nerve-** cells in the medulla oblongata, the pons Varolii, and other parts of the base of the encephalon, which **are** not employed in the transmission of sensitive impressions, or of the orders of the will to the muscles, and are endowed with the singular property of producing, even after a slight irritation, a persistent spasm in certain muscles, and especially in the neck."

Accompanying this rigidity of muscles from permanent spasm, there is supposed by this physiologist to be a similar excitement of the nerves going to the bloodvessels of the same muscles, by which the smooth muscular tissue of their walls is made inordinately to contract, depriving the affected muscles of an adequate supply of blood, and producing a rapid atrophy, while a similar irritation going to the voluntary muscles, secures their continued rigidity. It has been a mystery, how muscles could be actively contracting and yet rapidly wasting, and the hypothesis of Brown-Sequard affords a plausible explanation of this feature. He further says, p. 163: "That the paralysis of atrophied muscles is not the only cause of atrophy, is shown by the fact that this state (atrophy) of the muscles has often existed without paralysis, or at least before paralysis, and, sometimes, although there were convulsions in the muscles. Notta mentions three cases in which there were constant or frequent convulsions while atrophy was progressing."

Dr. Brown-Sequard holds the theory that the paralysis of the voluntary muscles may, in some cases, be itself produced by irritant contraction of the muscular substance of the capillaries supplying blood to the implicated nerves at their places of origin in the brain and spinal cord. Dr. S. Weir Mitchel, of Philadelphia, objects to this, that this explanation cannot be true, because the spasmodic contraction of the capillaries must relax. If it is admitted, however, that there may be permanent spasm of the voluntary muscles, it is not difficult to conceive the existence of such a state of muscles that are involuntary.

Dr. Mitchel's views are explained at considerable length, in the American Medical Times, July 9, 1864, and in the American Journal of Medical Sciences, January, 1865, p. 161, and in Gunshot Wounds and other Injuries of Nerves.

Dr. Brown-Sequard's theory finds further support, in the fact that in paralysis with a flaccid condition, atrophy can, to a considerable degree, be obviated by passive motion; and in this connection the following quotation has great interest:

Page 177. "Most of the morbid changes which have been attributed to paralysis do not belong to it, but are the result of irritation, either upon the nervous centres or upon the nerves, and the effects which are truly the consequences of paralysis, are due only in an indirect way to the absence of nervous action. Atrophy of muscles is chiefly due to a state of rest; changes of nutrition are chiefly due to dilatation of bloodvessels; ulceration upon the toes of animals, in which the nerves of limbs have been divided, only show the effects of rubbing the same parts upon the floor; ulceration and inflammation of the eye, after section of the trigeminal nerve, are chiefly due to physical causes (the drying of the cornea and conjunctiva, the prolonged action of light, etc.) All these may be, and have been, sometimes avoided.

"On the other hand, if we try to find out what is the power of cicatrization and repair, in cases of paralysis not complicated with irritation of nerves, we ascertain, as has long ago been done by Sir Benjamin Brodie, and as we have done since, by varying somewhat the mode of experimenting, that wounds, burns, and fractures may be cured as quickly in paralyzed parts as in others. If the influence of the nervous system is indirectly necessary to nutrition and secretion, it is, nevertheless, true that all the phenomena of nutrition and secretion may remain normal, when the action of the nervous system on the various tissues is missing."

The irritations which give rise to palsy of voluntary muscles, with the rigidity from irritative action of the involuntary nerves going to the same muscles, need not arise from irritating causes seated permanently at the origins of the nerves affected, but may be reflected from some peripheral seat of irritation. This affords not only ground of hope for relief, but a hint at the expedients of treatment.

Treatment.—Brown-Sequard thinks from his own experiments, and those of Matteucci, that in certain contractions of muscles, from irritations derived through the nerves from their centres, there is the condition of cramp. It is found that in contractions of this nature, the pain is increased by the elongation of muscles, and entirely relieved by their complete relaxation, either by the approximation of their attachments, or by tenotomy. He says (*loco citato*, p. 8): "Now, I have found that the greater is the resistance to the contraction of a muscle, the greater is the galvanic excitation that it gives to the nerves in contact with its tissue. On the contrary, if there is no resistance at all, as shown by Prof. Matteucci, after the section of the tendon, the galvanic excitation of nerves in contact with the contracting muscle no longer exists. It is not necessary for a muscle to contract in order to produce in nerves in contact with it a galvanic excitation. It is sufficient that it tends to contract." '

One of the experiments referred to, consists in preparing a leg of a frog, with its sciatic nerve lying upon the muscles of another limb. The latter muscles are galvanized, and the first muscles are excited through the sciatic nerve. It is found that when there is no resistance to the contraction of the latter muscles there is no galvanic excitation—no motion of the first muscles.

It follows that in cases of pain, in consequence of the contraction of muscles, as in cramps and rigidity, the division or extreme sudden extension of the tendons and muscles obviates the pain, by rendering null the galvanic excitation. This physiological explanation is also applied to explain the relief afforded by division of the sphincter in *fissura in ano*.

The same explanation also applies to a mode of treatment introduced by Dr. Wm. H. Van Buren, of New York, for fissure of the anus, which consists in forcibly distending the sphincter, by the separation of the two thumbs of the operator introduced into the anus. By this means the muscular fibres of the sphincter are either torn or so overstretched as to be temporarily paralyzed. It is claimed for this method that it has the advantages of a subcutaneous incision.*

* Contributions to Practical Surgery, by W. H. Van Buren, M.D., of Phila. Lippincott, 1865.

If the irritation causing the contraction were a permanent element, there would be little good obtained by dividing muscles and tendons; for, with the accommodation of the muscular fibres to their shortened relations, as the ends become again fixed the painful contraction would be reproduced. It is only when the irritation derived through the nerves is to some extent temporary, that permanent relief should be expected from this expedient.

In case of anal fissure, the ulcer may itself be the cause, by reflex action, of the painful contraction of the muscle. If the division of the muscle permits the ulcer to heal, the cause of

Fig. 6.
(From Brown-Sequard.)

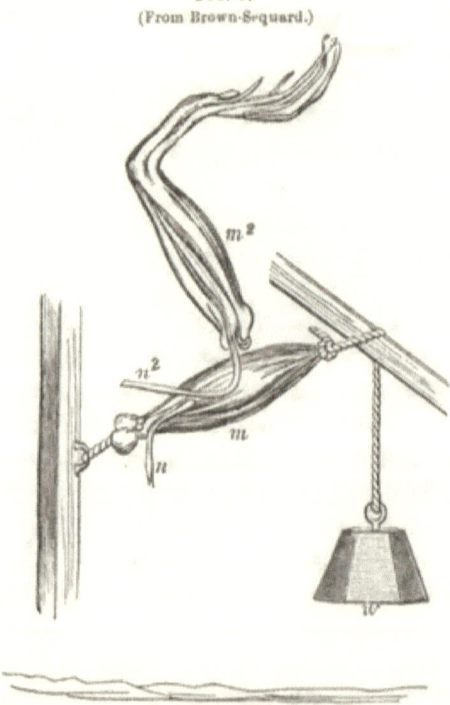

m A muscle made tense by the attachment of a weight w.
m² Another muscle with nerve n² lying upon the first muscle. When the muscle m tends to contract under the stimulus of galvanism, being restrained from movement by the weight w, the muscle m² contracts by galvanic excitation acting through its nerve n² which is distributed to the muscle m² of the other limb.

painful reflex contraction is removed; and after the granulating cicatrization has restored the connection between the divided portions of the sphincter, the muscle again performs its function without pain.

The rationale of treatment, here explained, throws a clearer light upon the benefits claimed by Guerin and others, to be derived from the division of muscles in distortions of the body and limbs, attended with a condition of the contracting muscles resembling a permanent spasm. This explanation may also lead to a clearer distinction between those cases which can be benefited by division of muscles and tendons and those which cannot.

In those disproportionately contracting muscles which escape the wasting process, a true hypertrophy occurs, and this continues after the irritating cause has disappeared. In these cases the division of muscles and tendons must hold out less promise; for, in a short time the wound must be cicatrized, and the intervening material must contract like cicatricial substance elsewhere, restoring the muscle to its original length and power.

The effect of the division of muscles and their tendons to interrupt their inordinate tonic contraction, mentioned with surprise by Stromeyer and Detmold, is now found to be equally well secured by extreme extension, which needs only to be employed for the briefest period and occasionally repeated; and in order to avoid all voluntary contraction, which would only afford so much more resistance to be overcome, the patient may be put into the state of anæsthesia.

Another plan of treatment may here be applied—that of extension, to the extent of exhausting and gradually lengthening the shortened muscle.

In considering the philosophy of the influence of extension in producing the elongation of muscles, the analogy of the paralyzing effect over distension of the muscular fibres of the uterus and of the bladder is in point. As long as the increased length of muscular fibres is maintained they manifest very little power of contraction. But the partial evacuation of the contents by the rupture of the membranes, and the escape of the amniotic fluid, or, in retention of urine, the free flow of the urine through a catheter, restores the muscular fibres to the conditions of activity, and they proceed in a little time to discharge their proper functions, or if maintained of a definite increased length they soon accommodate themselves to the new conditions, and acquire the capability of contraction and relaxation in the increased distance.

It is obvious then that in the treatment of deformities resulting from the contraction of muscles, the conditions of the muscular fibres of an over-distended uterus or bladder may be advantageously produced by artificial means. When the fibres have been drawn out, or induced to grow out of the desired length, the diminution of extension permits the muscular fibres to assume the capabilities acquired by elongated muscles after a dislocation. It is obvious, from these considerations, that while a very moderate force would only increase the power of the fibres in a given length, a greater degree of force would increase the length with temporary diminution of power. On the other hand, the extension of the shortening or shortened muscles diminishes the acting length of the yielding antagonist muscles, and permits them quickly to acquire increased power with diminished length. It is not necessary that they should acquire the same power as their overgrown antagonists, for we know that flexors and extensors are rarely balanced; and if it were the nature of the stronger muscles to overbalance the weaker, what grotesque shapes should we all very speedily assume.

It is not the true indication permanently to weaken the shortened or hypertrophied muscles, but to accustom them to normal positions and limitations of contraction; automatic and voluntary nervous action must be relied upon to control the tendency of unequal powers of unequal muscles to produce distortion. Until this habit is complete and lasting, the artificial restraint to the strong and support to the weak must be employed. It is from the premature discontinuance of the mechanical appliances to this end, that disappointment is too often due. In many cases it is necessary to resist, not only the tonicity of health, but the tendency to rigidity arising from nervous irritation; and, while this tendency lasts, the mechanical appliances must be worn. To obviate deformity is both wiser and more easy than removing it when it has once occurred.

For paralysis of muscles without tendency to rigidity, an obvious principle of treatment, is to secure such artificial movements as may most nearly imitate the natural. By this means, the circulation through the capillaries is kept free and nutrition may go on. Bones unite and wounds and burns heal in parts destitute of both motion and sensation if only the capillary circulation does not fail through motionlessness.

Where the cause of the paralysis is temporary, the great labor necessary to secure this artificial activity may meet its full reward in ultimate restoration of muscular power. It is not general exercise that is here so much needed, but "localized movements," which must be performed by a third person, or by some kind of mechanism, a good idea of which will be obtained by reading Dr. C. F. Taylor's "Movement Cure."

ELECTRICITY.—There never was greater discrepancy of opinion among practical men upon any subject, than exists among physicians upon the use of electricity in deteriorations of muscles, or perversions of their functions. It is suspected that this has grown out of the want of proper distinctions.

In proof of the power of electricity to preserve the volume of paralyzed muscles, nothing can be more conclusive than an experiment of Dr. John Reid, the account of which I quote from "Althaus on Medical Electricity," p. 143.

"Dr. Reid cut the spinal nerves in the lower part of the spinal canal in four frogs, so that both hinder extremities were insulated from their connection with the spinal cord. He then daily exercised the muscles of one of the paralyzed limbs, by a weak galvanic current, while the muscles of the other limb were allowed to remain quiet. This was continued for two months, and at the end of that time, the muscles of the galvanized limb retained their size and firmness, and contracted vigorously, while those of the quiescent limb had shrunk to at

least one-half their former bulk, and presented a marked contrast with those of the galvanized limb."

The form of electricity employed for the purpose of preventing
muscular atrophy, or stimulating the development of muscles,
when they are deprived of their appropriate nervous stimulus,
is the interrupted current of an *induction* apparatus, either
deriving its electricity from the decomposition of metals, as the
ordinary galvanic battery, or from the magnet, as in the mag-
neto-electric machines. The continuous current is found to
have no power in producing contraction of voluntary muscles,
though a slight contractile effort is produced upon smooth muscular substance, while these latter also respond more readily
to the interrupted than to the continuous current. In the language of Althaus (p. 177): "The contractions are observed,
not only at the commencement and the cessation of the current, as is the case with the voluntary muscles, but also while
the circuit remains closed. . . . . . . The movements induced in the muscular fibre-cells by galvanism, are not observed
simultaneously with the application of the electric stimulus,
as with voluntary muscles, but only a certain time after the
current has acted upon the tissue. . . . . . . Besides, the
motion once excited in the fibre-cells, continues a certain time
after the cessation of the current, and is not confined to those
parts to which the electric current has been directly applied,
as is the case with the voluntary muscles, but is propagated
to other parts of the same tract."

It is found that a continuous current of electricity, as well
as an interrupted current, secures in the non-striped or smooth
muscular fibres of the vessels, a contraction which lasts a considerable time after the cessation of the current. The effect is
to contract the bloodvessels which come in the course of the
current, diminishing the amount of blood which they contain,
and, consequently, diminishing the temperature.

Upon the voluntary or striped muscles, however, the contraction is immediate upon the beginning or cessation of the current, and ceases immediately and until a new excitation by the
beginning or cessation of a current. It follows, that the appli-

cation of electricity to a muscle produces two effects antagonistic to each other. First, a contraction, followed immediately by a relaxation of the muscular fibres, by which blood
is invited to the tissues, partly by the change of volume
increasing and diminishing in alternation, and partly by the
chemical change of the elements of the tissues, under the
physiological action of the muscles. By this increase of blood,
the temperature rises, and the nutrition and volume of the
muscles are maintained. Secondly, owing to the tendency of
the vessels at the same time to contract, a certain intensity is
given to the rapidity of the flow of blood, which resembles
that imparted by the presence of moderate cold. The continuous current acts very slightly upon voluntary muscles, but
quite considerably upon smooth muscular fibres, which give
contractility to the walls of the vessels. Therefore the interrupted current should always be employed to secure a better
nutrition of voluntary muscles.

In the state in which there is a rigidity or tonic contraction
of the muscles, owing to irritation of nervous centres, the
application of electricity would seem to be an expedient of
doubtful propriety, unless applied in shocks of such power as
to impair or destroy the irritability of the muscular fibres,
and thus occasion such a relaxation as may permit a free flow
of blood.

In the paralysis with relaxation, the current should be of
moderate intensity, and interrupted, to secure movements, and
the consequent circulation which movements induce, while in
paralysis with rigidity, the current should be of overpowering
intensity, and also interrupted, to paralyze the motor nerves
which secure the rigidity in the voluntary muscles, and also
to impair the contraction of the smooth fibres of the vessels
getting their supply of nerves from the sympathetic, so that
blood may flow more freely through the capillaries. By this
plan of applying electricity, a relaxation of the striped fibres
is secured, at the same time that the capillaries are relaxed and
enlarged, and for the time being, a better circulation is secured

through the rigid muscles, and an arrest or retardation of the wasting of muscular substance is realized. To avoid reflex irritation, there is an obvious propriety in making overpowering applications of electricity, while the patient is in the state of anæsthesia.

In the paralysis with relaxation, without diminution of temperature, there would be theoretic reasons **for** employing both continued and interrupted currents of moderate intensity. **The** interrupted, to exercise the muscles, and the continued current to give tonicity to the capillary circulation, **by the tonic effect of** a continuous current upon smooth muscular fibres. **In all** this, the chief object is to preserve the muscles from deterioration and prevent or remove deformity, while by other means, **or** by the processes of nature, the central cause of palsy **may** be removed, whether existing in the course of the nerves injured or diseased, or **in** the injured or diseased spinal cord or **brain.**

Brown-Sequard's theory of the action of electricity applied **to muscles affected with permanent contraction** is, that when applied in powerful shocks to **the affected muscles,** it exhausts the irritability of the muscular fibres, relieving them from the hold which the irritated nerves have upon them, and permitting them **thus to relax** at the same time that the smooth muscular fibres of the bloodvessels are relaxed **in** the same way, permitting a larger volume of blood to flow through the organs, and keeping up their nutrition until the spasmodic influences coming from the nervous centres can have time to disappear, by the restoration of **the** healthy functions of these centres, or by the **removal of external or centripetal** irritations which produce **their effects** by reflex action.

Brown-Sequard makes alternation of heat and cold, with the same **theoretic** explanation, in those conditions of parts in which sloughing occurs, especially upon the sacrum, in diseases and injuries of the spine, and in some forms of low fever. He explains that there is something more than inaction and pressure; that there is a contraction of the capillaries, which become relaxed by powerful interrupted electric currents, so that the usual amount of blood may again flow to preserve the nutrition of the parts and save them from sloughing.

Very much on the same principle, such parts are sponged with hot and cold water in rapid succession, the irritability of the muscular fibre-cells of the capillaries being supposed to be exhausted by the alternation, continued for a considerable time. Ice, or the coldest water, inclosed in a bladder, or in a rubber bag, is applied for about ten minutes, after which the bladder or bag is filled with warm water and applied, or if more convenient, a warm poultice is applied. Dr. Chapman, of London, has adopted this treatment with very varied therapeutic application.

Mechanical Treatment.—The indications for mechanical treatment of parts in paralyzed and spasmodic conditions are.

1. To maintain the normal relations against the distorting influence of weight, and of the faulty position in which the patient may prefer to keep them, and against the contraction of muscles which are not paralyzed, or only to an incomplete degree.

The means to these ends, are never-ceasing vigilance, directing the patient in his postures, and the employment of splints and bandages with elastic appliances to counteract whatever wrong tendency may be observed.

2. To protect partially paralyzed muscles from the annihilation, by overaction, of the little irritability which remains in them, and in spasmodic aptitudes, to restrain those movements, which, by reflex action, excite cramps or uncontrollable muscular action.

In palsy of the lower extremities, it is often important to protect the muscles from full action, and the limb from distortion, while yet the patient is permitted to move about upon his feet. The annexed cut (Fig. 7) affords a good illustration of the extent to which this can be done.

The joints are supplied with stops, for the purpose of limiting the movements of parts whose muscles are easily exhausted. The lateral deviations which are apt to occur, especially at the ankle, from want of muscular tension, are also prevented.

WRITERS' CRAMP (SCRIVENERS' PALSY) EXEMPLIFIES THESE PRINCIPLES.—This disease, arising from the overuse of the muscles moving the thumb and fingers, more often affects the muscles of the thumb and the index and middle fingers.

FIG. 7.
(From Bigg.)

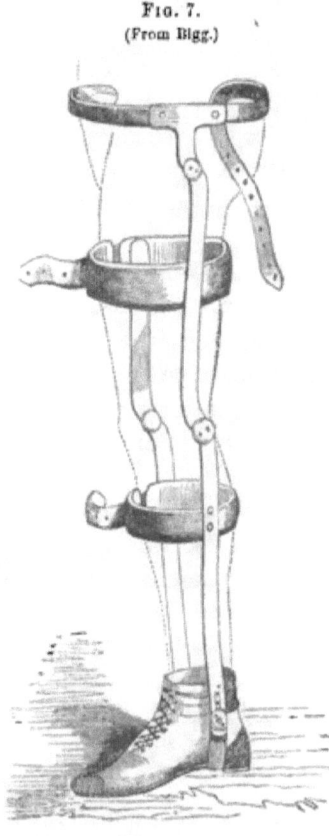

On attempting to write, the thumb is drawn into the palm of the hand, or the fingers grasp the pen with a spasmodic effort. Seamstresses, musicians, shoemakers, and compositors are often affected, incapacitating the sufferers from pursuing their usual avocations.

Mr. Bigg (Orthopraxy) recommends that writers should confine the affected parts, and educate the others to work up to them. Thus, if the thumb is affected with spasm, a gutta-percha investment is placed over it, which leaves only the last joint free. On the other hand, if the fingers are affected, similar restraints are placed upon them, and the thumb is left free. For the varied movements of music and sewing, such restraints are evidently inapplicable. Rest from the customary avocation becomes indispensable.

Velpeau advised for writers' palsy, that the pen should be held in a tube attached to the small end of a pear-shaped handle, just large enough to be grasped in the hollow of the hand. The movement of the pen is, of course, then made by the movement of the whole hand, aided, in a slight degree, by the fore and middle fingers.

3. To keep up circulation and consequent nutrition, in order that when the cause of impaired or perverted innervation is removed, the muscles may be in good condition to resume their functions.

Fig. 8.

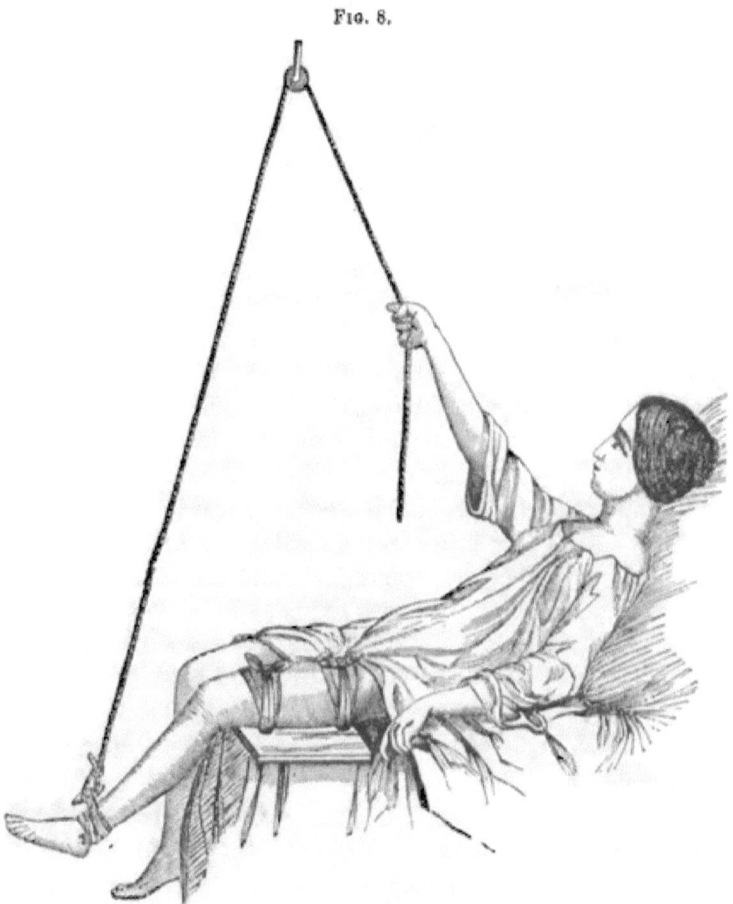

Plan for exercising a palsied limb (from Bonnet).

The mechanical appliances for this purpose are as varied as ingenuity can suggest, but it is important to give the patient such machinery as will afford him employment at the same time that the affected part gets its proper exercise. A hint at a very simple method of doing this will be afforded by the figure on the preceding page, from Bonnet (Fig. 8).

STRABISMUS.—Deviations of the antero-posterior axes of the eyes from their normal parallelism, may be the result of a spontaneous effort to bring into the best position for vision, the most transparent part of a partially opaque cornea; or to turn away from the direct line of vision, the cornea of an eye having imperfect vision from opacity of the cornea, or of the lens; from imperfection or disease of the retina, or from irregularity in the refraction of light with resulting distortion of images. In this class of deviations any attempt at treatment which does not remove the organic defects on account of which the abnormal direction has been assumed, must end in disappointment.

In the more numerous class of cases arising from reflex action, or possibly in some instances, from irritation, either in the course of the nerves or at their origins, the primary treatment has reference to the removal of the irritation.

It is possible that strabismus occurs in some instances without any pathological cause, as the result of the attempt to give the pupils maldirections, simply for drollery.

Once confirmed, the deformity appeals for aid and finds an imperfect correction in the division of the shortened muscles. Only the internal and external recti are usually divided, respectively for inversion and eversion of the pupil.

It is much easier to make an open vertical incision over the insertion of the muscle, pass a blunt hook under it and cut it with blunt-pointed scissors; but the extensive division of the fascia sometimes leads to a deviation in the opposite direction. On this account, and to keep the natural fulness at the inner canthus, it is now most customary to make a longitudinal incision, pass a hook under the muscle, insinuate the scissors under the mucous membrane and cut in the dark.

The frequent necessity for searching again, and in some cases for repeating the operation after an interval, compensates for exemption from the risk of over-correction, and the avoidance of a sunken appearance at the inner canthus. In diverging strabismus this consideration does not apply. In aggravated cases · of strabismus, the division of the contracted muscles is usually satisfactory in its result, because, if it fails of complete restoration, it is at least an improvement on the previous condition, and the danger of turning the pupil in the opposite direction is very slight.

Over-correction has so often occurred, however, in spite of the greatest care, and the failures arising from a timid mode of operating, in view of the danger of doing too much, have been so frequent, as greatly to reduce the favor in which this operation was held soon after its introduction by Stromeyer and Dieffenbach in 1838 and 1839.

The success recently attending attempts to change the contracting distances of the internal and external recti muscles, renders it probable that the expedient of dividing the muscles will be reserved for those cases which resist the treatment by *exercise.*

Exercise of the muscles of each eye separately, in order to give the muscles varied capabilities, elongating shortened muscles, and contracting lengthened ones.

This is most conveniently done by placing before the eyes a pasteboard blind with large orifices opposite each eye. Across each of these orifices a slide, having a smaller orifice in it, is so placed that it may be readily moved from or towards the centre. Where only one eye is ordinarily employed in vision, the orifice over the best eye is entirely closed and the slide over the other is so moved that the poorest eye in convergent strabismus can only see by an extreme abduction. The exercise in reading and looking at various objects is continued, until the discomfort becomes intolerable, and repeated several times a day until the contracted muscles acquire the power of elongation, and the elongated muscles the capability of contracting to a shorter length. It is difficult, if not impracticable, by this expedient to exercise the two eyes at the same time.

STEREOSCOPE.—In the "Philadelphia Medical News and Library," for November, 1864, copied from the "London Medical Times and Gazette," September 24, 1864, is a very brief reference to some experiments of M. Javal, of Paris, related in a Congress of Ophthalmologists at Heidelberg, 1864, *in the use of the Stereoscope in correcting convergent squint.* The method consists in placing upon a card, cut of the ordinary size of those upon which stereoscopic pictures are placed, two slides (like a slip of paper around a package of envelopes) upon each of which a wafer is attached, both being at the same height upon the slides. This card, thus supplied with its wafers, is then placed in the position the picture is intended to occupy. For convergent strabismus the slides should be separated until two slides and two wafers are seen. Then by steadily looking through the prisms, the pupils diverge and the two images fuse into one. By the prolonged and frequently repeated exercise of the muscles of the eyes in this manner the parallelism is expected gradually to be regained.

PLANE PRISMS.—In Braithwaite's Retrospect, No. 50, January, 1865, Am. Ed., p. 158, is an article by Ernest Hart, Ophthalmic Surgeon to St. Mary's Hospital, London, copied from the Lancet, July 30, 1864, in which he gives an account of *the use of plane prisms* in correcting the deviation of the eyes in strabismus. He relates cases of success, and explains the result as owing to the effort to change the direction of the affected eye to avoid the double images which the prisms tend to produce. The principle is that of the stereoscope, which causes two images to appear from a single central object, and a single image to appear from two objects placed at the proper distance from each other, and seen with both eyes at the same time. It is the effort of the muscles of the eyes so to direct the pupils that these double images may be avoided, which accustoms the shortened muscle to act in longer distances, and the elongated muscle to act in shorter distances, and which by degrees secures a restoration of the proper parallelism of the eyes. The bases of the prisms are placed opposite the direction of the squint. So, for convergent squint or cross-eye the prisms would be arranged as they are in the stereoscope, and in divergent squint in the opposite direction. The angles of the prisms are made to vary according to the effects desired, using those of 4, 6, 8, 10, and 12 degrees, employing for constant use those of the degrees which only double the images at considerable distances.

Cases which are extreme, are to be treated by division of the shortened muscles, and those which are sympathetic of still existing irritation or inflammation are to be reserved for treatment after the central or reflex irritation has subsided.

The treatment of strabismus without operation is in its experimental period, and all that is here attempted is to state the progress of experiment.

III. INFLAMMATION AND PERVERSIONS OF NUTRITION INJURING OR DESTROYING THE TISSUES, PERVERTING OR NULLIFYING MUSCULAR CONTRACTION, SOFTENING BONES, DISTORTING AND STIFFENING JOINTS, AND PERVERTING, IMPAIRING, OR ABOLISHING THE FUNCTIONAL MOVEMENTS OF THE PARTS AFFECTED.

Whether the inflammatory diseases which produce these results arise from wounds or injuries of the joints, or occur as the localizing of some more or less permanent constitutional condition, the indications for treatment and the ultimate results are very similar.

The inflammations from accidents are, usually, at first, synovial, attended in the beginning by excessive and thinned secretion of synovia, and afterwards by suppuration, imposing upon the nervous and vascular systems violent perturbations, and involving very soon, the surrounding tissues in inflammation, with softening and ulceration of the non-vascular articular cartilages. The processes are liable to be more rapid than is usual with spontaneous diseases of the same parts, with an earlier cessation of diseased action, and the establishment of permanently altered conditions. Where, by appropriate treatment, the inflammation is aborted, before these grave results accrue, the course of the disease may be reduced to a very short period.

Spontaneous synovitis may be of two kinds—rheumatic and strumous—each more or less simple or complicated with inflammation of surrounding tissues. The rheumatic inflammation is the more painful and variable in severity, with exudations of solidifying material, which may afterwards seriously diminish the mobility of the joint affected, in consequence of roughness of surfaces, adhesions, the production of movable bodies, and the diminished flexibility of surrounding parts, the articular cartilages in some cases ulcerating, or after fatty degeneration, becoming disintegrated under the friction of the opposed cartilaginous surfaces, until the rough bony surfaces are brought into contact, to end ultimately in eburnation with surfaces polished by friction, or in anchylosis. The constitutional irritation may be such as to require exsection or amputation before the natural termination of the disease. Where several joints are simultaneously attacked with the most acute form of this inflammation, the patient's life may be destroyed very suddenly, by the depressing influence of nervous shock.

The strumous variety of synovitis may present all grades of severity and destructiveness, and be at first, accordingly, painless or more or less painful. The easy expansibility of the synovial membrane saves it from painfulness when the changes of volume are very slow, and the involvement of the articular cartilages in destructive processes can hardly add to the pain, as they are destitute of nerves, and must, therefore, be destitute of sensibility.

Whether we accept the anatomy of the synovial membrane as a tube terminating at the articular cartilages, or as a closed sac, there is no practical difference, for if the articular surface has a synovial covering, it is destitute of vessels and nerves like the cartilage beneath. This is the reason why a slow inflammation, involving destruction of parts to a frightful extent, may occasion very little pain, as long as the bony tissue escapes the inflammatory process.

Inflammation, commencing in the spongy heads of the bones, must be supposed to be attended with more pain and lameness, in consequence of the free supply of nerves, which are exquisitely sensitive in **the** inflamed **state.** So slow is the process in some cases, that the nerves accommodate themselves **to** the slowly changing volume, remaining surprisingly free from painful sensations. From this it follows that, in a great **number** of cases, the diagnosis **must** be exceedingly obscure as to whether the soft **parts or the bones** have been originally attacked. **It is fortunate that** this distinction is of no practical importance, as the treatment of one location of disease is equally **appropriate** to the other. This distinction, however, may be considered clear, that when, in the course of a joint inflammation, there comes a great aggravation of painfulness, with spasmodic muscular contractions, especially aggravated by motion of the joint, or pressure of its articular surfaces against each other, the bony tissue has become involved in inflammation, and, perhaps, uncovered by the disintegration of its investing cartilage, so **as to expose its** rough surfaces to friction.

After the subsidence **of the acute symptoms in strumous** synovitis, and in **the earlier** stages, **in cases in which the ini-** tiation of the disease is **by slow** and doubtful steps, the pain is of the character of *aching*, increased by the dependent position. Perversions of sensation sometimes exist, as of cold or heat, when there is really no change of temperature.

In the second stage, after weeks or months, the pain is *gnawing*, with soreness referred to the bone and starting of the limb at night. In this connection, it is proper to quote the emphatic language of Barwell, "On the Joints," American edition, p. 136, where he says:

"These sensations have been supposed to be caused by an

ulcerating process going on in the cartilage, and so indeed they are, but only in a secondary manner, for the pain is directly produced by the hyperæmia of the subarticular vessels of the bone produced by the cartilaginous inflammation. Cartilage, whether healthy or diseased, possesses no nerves, the only conductors of impressions to the brain; then it is as insensible when inflamed as when not inflamed; but hyperæmia produced thereby, in such an unyielding structure as bone, sets up these painful symptoms."

"Another sensation, attributed with equal want of precision to ulceration of cartilages, is tenderness on pressing joint surfaces together. The origin of this symptom, although obscure, I believe myself to have detected. By questioning, for years past, every patient that came in my way; by observing the species and succession of different sensations, and examining, when possible, the joints of those whose symptoms have been thus noted, I have come to the conclusion that this tenderness indicates that the articular lamella has given way over a larger or smaller extent, and that the cancelli are laid bare to the joint."

Again, p. 244: "Pathology shows us that in a synovitic disease no special action is produced among the muscles, until the bone *underlying the cartilage* becomes affected. Again, we see that when *that portion* of bone is *primarily* diseased, the spasms of the muscles producing the start and shock are among the earliest symptoms. We find that a carious state of this portion of bone is extremely irritating, and sets up not merely temporary spasms, which pass like electric shocks over the limb, but that a slower and lasting contraction takes place. This phenomenon affects nearly all the muscles moving the lower bone of a diseased joint, but it predominates in the flexors, and, therefore, the lower bone becomes rigidly bent upon the upper; the muscles feeling tight and cord-like under the skin. Such contraction is produced by a morbid form of reflex action, carried from the nerves supplying the part to the muscles. . . . . . . This contraction continues during sleep, and is of greater power and duration than any voluntary contraction. The spasms are more violent when the will is withdrawn, and they precede the paroxysms of continued pain; and the muscles **affected** with this peculiar contraction waste with more rapidity than in any other disease, except in certain cases of irritation of the spinal cord producing spasmodic muscular contraction. . . . . . Although the muscular phenomena are originally produced by the irritation of the joint disease, they eventually increase, or altogether support its morbid actions, **by** forcing one tender bone surface against another. . . . . . . Sometimes, but at a later stage, when the tonic contraction of the muscles produces dislocation, **the** spasms and **starts abate very** much indeed, or disappear altogether—the **displacement of one bone upon the other giving** instant **relief—a proof in aid of the fact**" (inference) "**that it is the mutual pressure which produces the whole train of symptoms.**"

Another class of joint destruction arises from immobility of the parts with the ordinary pressure occasioned by the tonic contraction of the muscles.

M. Bonnet, of **Lyons,**[*] **quoted by Dr. H. G. Davis, in** Transactions of the American **Medical** Association, **1863,** gives the result of observations by himself and Tessier, **on the** effects **of prolonged** immobility **of joints. He says, "I am about to demonstrate,** anatomically, **that long-continued immobility can produce severe disease in healthy joints,**" and then goes on to give the following results:

1st. Effusion of blood and serum in the articular cavities.

2d. Injection of the synovial membrane, and the formation of false membranes.

3d. Alteration of the cartilages.

4th. Anchylosis.

He says, "I have not mentioned stiffness of the joints, as among the anatomical lesions which immobility produces. This stiffness is frequently observed, and ought to be particularly considered as an effect of the alterations which the autopsy reveals in the cartilages and in the synovial membranes."

* Traité des Maladies des Articulations, accompagné d'un Atlas, avec 16 planches, par A. Bonnet, Prof., etc., etc. Paris et Lyons. 1845. Tome I, f. 9, et seq.

M. Tessier has the credit of having first noticed the effects of immobility upon the joints. The latter says, "I have almost invariably found in all the articular cavities of the diseased limb, even in those most remote from the solution of continuity, the secretion of synovia replaced by bloody serum, and even by liquid blood almost without admixture. In one case, and one only, I have found clots of blood. This was in an old man, confined six months for fracture of the neck of the femur."

Bonnet says sanguineous effusion and injection of the synovial membranes are the two first effects produced by immobility. In all the cases in which M. Tessier observed them, they were already supplied with vessels, and adherent to the cartilaginous surfaces.

"Their existence appeared to demonstrate that long-continued repose can produce, in the joints, lesions of an inflammatory nature."

"Continued immobility with pressure can produce redness, swelling, softening, erosion, and wasting away of cartilage. The redness may be uniform or punctate. Where the cartilage is not eroded, it presents itself under the form of ecchymosis, more or less deep. On the contrary, where the cartilage is eroded, it is unequal, dotted."

Dr. Davis, in the same paper, p. 158, quotes from Dr. Willard Parker of New York: "I have often seen in the knee-joint, after amputation, when the joint was opened, that when the surfaces had remained long in contact, the synovial membrane and cartilage were removed by absorption, and the bone, at the same point, dead for from an eighth to half an inch in depth.

"In exsection of the knee-joint, on opening the cavity, I have found the same destruction to have occurred.

"The same pathological condition is observed in the hip-joint. Indeed, I regard it as established, that if the surfaces of joints are allowed to remain long in a fixed position, the pressure from the muscles causes destruction of the substance making the wall of the joint. We see the same condition resulting in a joint, that happens when pressure is allowed upon the heel in the management of a fracture, viz., ulceration and sloughing."

Dr. Wm. H. Van Buren, of date 1860, in a letter to Dr. Davis, says,. "In thinking over the many cases of diseased joints which I have examined after amputation, and otherwise, my impression is, that the greatest amount of disorganization has existed when the opposed surfaces of the joints have pressed against each other. . . . . . . A recent case, in which I exsected the knee-joint of a young woman, for chronic strumous synovitis, afforded strong and indisputable evidence on the point in question. In the centre of each of the articular depressions of the tibia, I found a plate of necrosed bone, each about the size and thickness of a dime, lying loosely upon a bed of granulations. The articular surfaces were also profoundly altered, but except at these points, not beyond the possibility of repair to the extent of anchylosis."

In this connection, it is proper to say that, through the politeness of Prof. Edmund Andrews, of Chicago, I have been shown a specimen of inflammation of the knee-joint, exsected by this gentleman, in which there was an exfoliation, half an inch in diameter, upon one of the condyles of the femur, and a corresponding necrosis, preparatory to exfoliation, upon the opposed surface of the tibia. These were at points where the erect posture or the straight position would bring the greatest pressure.

The accompanying cuts from Dr. Andrews, Figs. 9 and 10, illustrate this state of things. In the hip-joint, the necrosis is seen to be both upon the head of the femur and the upper portion of the acetabulum. In the knee-joint necrosis is most likely to occur where the opposite surfaces of the femur and

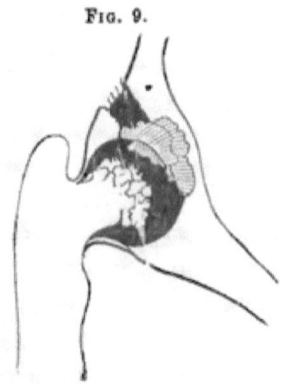

FIG. 9.

Sectional view of a case of **Hip-Disease.**

of the tibia receive the greatest pressure, and where the patella rests upon the femur.

Fɪɢ. 10.

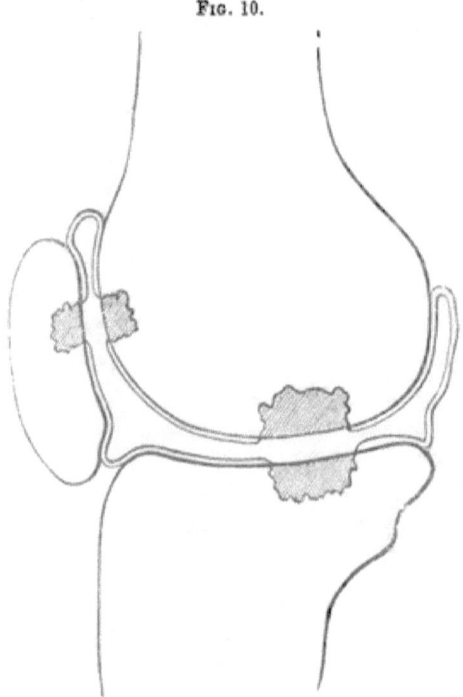

Vertical Section of a Knee-Joint, showing that Necrosis in one surface is attended by Necrosis in the surface in contact with it.

It has been claimed that the occurrence of necrosis upon a joint surface, induced a similar disease in the corresponding point of the surface of the opposite bone. The light in which the subject is here viewed, renders it probable that pressure is the cause of necrosis in both bones, and that the necrosed portions are opposite each other, not because one produces the other, but because this situation is necessary to the mutual pressure of the joint surfaces, and the observed fact with some exceptions is that the exfoliation generally takes place where the shape of the bones renders the pressure the greatest.

Dr. Davis quotes two cases from Bonnet, of fracture of the neck of the femur within the capsule, in which (the patients dying after long confinement) all the joints below the injured joint were in a state of disorganizing inflammation, while in the joints injured, no such result was found.   Dr. Davis explains the mystery thus: "The head of the femur broken off and lying loose, does not feel the force of the muscular pull; the other articulating surfaces do."

A painful inflammation of joints, however excited, at length induces cramps or spasms of the muscles passing the affected joint, which may be both paroxysmal and permanent, sometimes acting with painful exaggeration, but pulling constantly, with an increase of the natural tonicity; the stronger flexors usually gaining upon the weaker extensors, until the joint becomes so flexed as to relieve the inflamed joint surfaces from the pressure which had been at first occasioned by the muscular contraction.

The effect of this permanent spasm of the shortening muscles, is to produce a rigid state, to which Barwell has recently given a new name, that of "contracture."   This is his own definition of the condition.   "On the Joints," p. 315:

"The muscles which are affected with contraction, gradually shorten organically and permanently; they become passively *contractured*, that is to say, their decrease in length is not merely a passing state, which will disappear when the stimulus ceases.   They become fixed in this shortened condition, either by the gluing together of their elements, or some like cause, and they cannot of themselves resume a relaxed and lengthened position. . . . . . It appears to me that the change is located in the sheath of the fibres, rather than in the fibres themselves.   Every fibre of a muscle is composed of a sarcos and of an investing wall; the active contraction of a muscle is produced by the shortening of the flesh; passive *contracture* supervenes after the interior has been for some long time in this shortened condition, when the investing part adapts itself permanently to that shape, and each wall of every muscle-cell is fixed in its abbreviated form.   Moreover, each portion of areolar tissue investing the fibrous bundles, assumes permanently, the new form impressed upon it by the inclosed contracted sarcos. Such change does not forbid continuing active contraction, for the state, *contracture*, depends upon the change in the passive parts of the organ, to which ordinary muscular action may be added."

Treatment.—In the first stages of any disease of a joint involving the synovial membrane and articular cartilages, the first indication, obviously, is to give the parts the conditions of the least possible excitement or irritation. Rest, and such moderate extension as to counteract muscular contraction, measured by the feeling of comfort, meet the requirement. Warm or cold applications, according to the amount of heat to be overcome, and in violent inflammations, with rapidly accumulating pus within the synovial membrane, free, that is, long incisions into the cavity of the joint should be made, to give free outlet to the offending fluid. One of the valuable contributions to surgical science, by the late lamented Dr. E. S. Cooper, of San Francisco, was a more clear and emphatic enforcement of this expedient in acute inflammation of joints, from disease and injuries. The incisions should be free enough to discharge not only pus, but coagula, concretions and sloughs whose presence, in a decomposing state, poison the parts, aggravating the inflammation, and increasing the amount of destruction. After free incisions, it is practicable to wash out the materials that do not readily flow of their own accord. The old-fashioned way of "lancing" a joint is altogether inadequate. It is hardly necessary, however, to say that the heroic incision need not be practised in anticipation of the active and destructive inflammation, but only after the unequivocal establishment of suppuration; for, by the proper sealing of the wound, absolute quietude, arterial sedatives, and regulation of the local temperature, with the necessary depletion by the alimentary canal, the dreaded inflammation may often be prevented. Later in the case, local stimulants, as mercurial ointment, liniments, affusion of water, and passive exercise, and then active exercise practised during sufficiently brief periods, close the treatment.

It is proper to state that in the rheumatic variety of inflam-

mation, there may be an extent of swelling, and a distinctness of fluctuation, very likely to mislead a careless observer
with the supposition of the existence of pus, requiring to be
evacuated. The gelatinous character of the rheumatic exudation should theoretically give rise to an elasticity rather than
a fluctuation; but, through the thickened condition of the investments of the joint, it may be impossible to determine by
any manual examination, whether there is pus or rheumatic
exudation. It is, in these cases, that the grooved exploring
needle may be a very welcome aid in the diagnosis.

It is not proposed to discuss the distinctions by which the
question of the existence of rheumatism is to be determined;
but, if other joints have been affected in succession, or the
same joint has been attacked on previous occasions with rapid
improvement after violent local symptoms; or, if there has
been a frequent recurrence of attacks of a milder grade, it is
safe to say that the case is one of rheumatism.

The disease may, however, be no less destructive on account
of its being rheumatic, nor is there any difference required in
the local treatment. The constitutional treatment will generally require to be less antiphlogistic than in cases of traumatic inflammation, and if an alkaline treatment is proposed
to be adopted, it would be well to be certain that the case is
not one of a strumous character.

If, in the chronic state, in case of enlargement, the exploring needle reveals the existence of a thin fluid, there can
be no harm in permitting a portion to flow out by the side of
the needle, or through a minute trocar, after which, if an acute
inflammation should follow the introduction of a weak iodine
solution, benefit may occur in the end, from an increased disposition of the synovial surfaces to absorb the liquid contents.

The *constitutional* treatment cannot differ from that of diseases of other parts presenting the same pathology. The
principles are pretty well explained by Abernethy, in his little
work, entitled " Surgical Observations on the Constitutional
Origin and Treatment of Local Diseases." The chief advance
upon Abernethy, is in the employment of iodine, a great

amount of which, however, is thrown away, through the neglect of that eliminative treatment, without which the correcting effects of iodine and its compounds will too often fail to be realized.

Barwell, in his work, " On the Joints," reproduces Aberne-thy's principles, and, in addition to them, he makes a very important distinction between the treatment appropriate to those cases of strumous diseases in which the blood fills the capillaries, producing a considerable degree of redness of the general surface, and those cases in which there is pallor and emaciation. The former cases require purgatives before iodine, iron, or other alterants and tonics, will produce any abatement of the local diseased activity. It is often surprising how rapidly a local disease will ameliorate, after a thorough pur-gative, acting upon all parts of the intestinal tube and tribu-tary glands, to remove secretions and excrements too long retained. The repetition once a week, for a child five or six years old, of a combination of a grain of calomel, a twelfth of tartar emetic, and half a grain of compound extract of col-ocynth, or some similar purge, will often work wonders, after an entire failure in the employment of iodide of potassium, iodide of iron, quinia, and cod liver oil, without this purgative to precede and interlude the alterant and tonic treatment.

In inflammations of the more acute kind, much is gained in the early periods by securing, temporarily, a decided control over the circulation; and for this purpose, no agent, yet known, is equal to veratrum viride. During the depression of arterial action the capillaries regain much of their contrac-tility.

Blisters, setons, issues, moxas, and cauteries, actual and potential, if used at all, should be employed in the later periods of the disease. They have all fallen into great neglect of late, apparently from the present fashionable practice of directing sugared medicine. The pain of the application of moxas and cauteries may be altogether neutralized by ether or chloroform, but still they are in great disfavor. It is difficult to conceive that the confidence reposed in them half a century ago was altogether a mistake, though it was, perhaps, too great and too indiscriminate. There should not only be a freedom from activity of inflammation, but the location should be far enough from the seat of disease not to subject the near branches of the same nerve to irritation. The power of burns and wounds, and of diarrhœa from irritability of the mucous membrane of the alimentary canal, to produce those palsies, both of motion and sensation, now denominated reflex, is a presumption in favor of the power of counter-irritants to change or allay the painful excitement of diseased joints, and attendant muscular spasm. To discriminate the cases in which to employ the agency is the point. Perhaps the most powerful and least painful of all these, is the actual cautery, applied very rapidly and very lightly, and frequently repeated. And yet the agent has gone into almost entire disuse.

Barwell quotes a case from Rust, of Vienna, of a young gentleman suffering severely from hip disease, whose parents could only persuade him to undergo the application of the hot iron by promising to take him to the theatre that evening. The application was freely made, and the boy's pains were so much diminished thereby, and he was so cheerful, that he insisted on the fulfilment of the promise, and he greatly enjoyed the entertainment.

The influence of the counter-irritant on the disease may perhaps have been overestimated, by its power in paralyzing the spasmodically excited muscles, and thus diminishing the pain of their contraction.

There comes a period in the process of suppuration, when the pus has escaped from the joint, in which the propriety of opening the cavity and discharging the contents becomes a question.

I incline to the advice of South, who says, in speaking of abscess of the hip-joint (South's Chelius, vol. i, p. 299), " On the whole, I think it preferable not to meddle with abscesses of the hip-joint, unless they excite much constitutional irritation, and until the skin is on the point of ulcerating; then they may be punctured, and untoward symptoms rarely follow."

The ultimate treatment, after the subsidence of inflammatory action and the cicatrization of ulcerated parts, may **require** incisions and the employment of apparatus, in order to sunder adhesions and **remove** distortions.

EXTENSION.—It is appropriate here to notice, more in detail, the history and philosophy of extension, **as an** element of treatment in inflamed joints. The earliest account of the application of extension during the period of inflammation, which I can find, is by Brodie in his Pathological and Surgical Observations on Diseases of the Joints, p. **145 (published about 40 years ago)**, and quoted in South's Chelius, American edition, **vol. i, p. 296**, where, speaking of disease of the hip-joint, he says: " At a later period, when, in consequence of extensive destruction of **the** articulation, the muscles begin to cause a shortening or retraction of the limb, I have found great advantage to arise from the constant application of an extending force, operating in such a **manner as to counteract** the action of the muscles. **For** this purpose, an upright piece of wood may be fixed to the foot of the bedstead, opposite the diseased limb, having a **pulley at** the upper part. **A bandage may be placed around** the thigh above the condyle, with a cord attached to **it,** passing over the pulley and supporting a weight at its other extremity. I will not say that the effect of such a contrivance is to prevent the shortening of the limb altogether, but I am satisfied that it will, in a number of instances, render it less than it would have **been** otherwise, at the same time preventing, or very much diminishing, that excessive aggravation of the patient's sufferings, with which the shortening of the limb is usually accompanied."

This quotation should **be regarded as** remarkable, on account of the length of time the suggestion of extension seems to have **lain** dormant. **It is** remarkable that Brodie himself, finding the expedient effective in diminishing the painful contrac**tions of** the muscles in the later stages of the disease, did not think of using it to prevent the conditions upon which these painful contractions depend.

Dr. Lionel J. Beale, in his work on Deformities, published soon after Brodie's Observations, speaking of contractions of the knee, p. 99, says: "Gradual extension, friction, the application of vapor, and mechanical aids to allow of exercise, are the means to be employed. *But before having recourse to them, we must be certain that all active mischief has ceased, and be very cautious that we do not reproduce it.*"

It is plain that Beale did not understand the expedient of quieting the irritation of inflamed joints by extension.

In the London Lancet, vol. i, 1852, p. 115, J. Cooper Foster, F. R. C. S., M.B., &c., says that Mr. Key treated two cases of hip disease with a straight splint, while Mr. Foster was a dresser for him. Nothing, however, is said of extension, though this must have been subsequent to the cases treated by Brodie, by extension. Several cases are quoted from Mr. Foster's practice, in which recovery occurred with limbs in proper position with preservation of motion.

So it seems that at the time Mr. Foster wrote, extension had not come into general recognition.

Coming down to 1860, we find Barwell, in England, in his work "On the Joints," p. 144 (speaking of local treatment), saying: "The first and most important part of the local treatment, is rest. A time arrives, as we shall see, when it becomes a grave question whether entire immobility should or should not be continued, but there can be no doubt that at first the joint should be kept perfectly still by bandaging a well-padded splint upon the limb. The joint itself must be left uncovered by the bandage, for the application of any remedies that may be desirable, which in this stage of the disease belong chiefly to the class counter-irritants and derivatives."

At this date, then, Barwell applied splints in the earlier stages of diseased joints, to secure rest, and not to produce extension and separation of adjoining surfaces, with reference to the avoidance of the irritation of the pressure of inflamed surfaces by muscular contraction.

In the United States, Dr. Alden March, of Albany, reasoning upon his own experience, and upon the remark of Sir

Charles Bell, that "the thigh of the affected side is thrown over the other, that the head of the bone may be raised so as to relieve the inflamed socket," came to apply continuous extension after a resort to forcible extension in the later period of a case of hip disease in 1852.

Dr. March's splint was carved of wood, made as light as possible, and wide enough at the top to inclose half the circumference of the thigh and pelvis, with an open orifice opposite the trochanter to avoid pressure there. The ordinary perineal band was attached to this upper end, and extension was made upon the leg and ankle, holding the foot against a foot-piece placed at right angles to the splint.

Dr. March quotes Dr. Physic as having employed his long splint in hip disease, and Dr. William Harris, of Philadelphia, as having reported in the Medical Examiner, for January 19, 1839, cases of hip disease treated by extension.*

Dr. March's apparatus was an approach toward a portable splint.

Before this, or about the same time, Dr. H. G. Davis, of New York, constructed a much more portable apparatus with elastic rubber extension, some account of which was published in the American Medical Monthly in 1856, with the announcement of these indications for mechanical therapeutics :

"1. That in all diseases of joints without the capsular ligaments, extension should be applied when the joint becomes functionally immovable.

"2. In diseases within the capsular ligament, the extension should be applied at once.

"3. In immobility, from whatever cause, the limb must either be frequently moved, or it must be extended."

In Transactions of the American Medical Association for 1863, p. 150, Dr. Davis gives wood-cut illustrations of his apparatus, and says : " I insisted from the first, on the fact" (principle) "that mobility is natural to, and required by, a diseased, as well as a healthy joint, and introduced that as one of the principles of my treatment.

* See Trans. American Med. Association, p. 479, for 1857.

"I insisted, also, from the first, upon the fact" (principle) "that pressure, mostly owing to muscular contraction, is the most active agent of destruction in the morbid process, which it is the object of my treatment to overcome, and I therefore directed my efforts to obviating, *in all the stages of the morbid process*, the pressure to which the parts diseased are exposed. To attention to this, I ascribe my main success.

"The distinctive principle of my treatment, is the procuring to the diseased structures support without pressure and motion without friction.

"The **treatment itself**, concisely stated, consists in abstraction of the joint affected, by continued elastic extension."

There comes very soon, in violent inflammation, a preponderance in **the pull** of the flexors over that of the extensors, **as if for the purpose of** placing the joint surfaces in **such relations** to each other as to relieve them from pressure. **The** good result of Nature's therapy is secured by a position of the limb which renders it afterwards useless, unless a surgical process is resorted to, for the purpose of restoring the limb to the straight position. I cannot enforce this point better, than by quoting from Tamplin "On Deformities," American edition, 1846, p. 168, this clear statement:

"During inflammation (of synovial membranes) there is a constant effort to keep the joint at rest, the slightest motion occasioning the most severe pain; **and the** position in which there is the least amount of pressure **on the** articular surfaces, is undoubtedly the flexed position. Be this pressure ever so slight, **it** must increase the inflammation in **the** synovial membrane, and consequently, **the** pain, which **is** at times most acute, **requiring the** most active* measures for relief. The flexors, therefore, are constantly acting, and they become eventually **contracted, from the flexed position** being maintained during the inflammatory attack."

This pathology should have suggested extension, but it did not. Tamplin accepts the anatomy of a synovial membrane over the articular cartilage, with sensitive nerves in it, which is a mistake. The pain from pressure at a period so early, is

5

probably from the influence of the pressure upon the inflamed bone immediately subjacent, in which the nerves sensitive to pain have been aroused into activity.

The force to be applied, **and** the pain to the patient, is **far** less to prevent this deformity, at the same time that the surfaces are relieved from pressure upon each other, than that which is required to overcome the deformity, when once it has occurred.

Extension in inflamed joints, so far from being painful, is, on the other hand, positively grateful to **the patient ;** so that the sensation **of comfort** may safely **be made the test of** the amount of extension required.

An instance, illustrating this, occurred to the **writer, in the** case of an acute traumatic inflammation of the knee-joint, with extensive suppuration and burrowing of pus along the course of the muscles. This patient came under my care, at the end of the sixth week from the time of the accident, with a flexure **of** the knee to an angle **of 45** degrees. **Adhesive** plasters were placed upon **the leg, as** in the treatment of oblique fracture of the femur, and extension made **by a twisted rope passing from** the loop of plaster **under the** sole **of the foot, to** a point eighteen inches beyond, **while counter-extension was made** from **the** chest, by the attachment of long **and wide** adhesive slips, before and behind the chest. Extension proved not only a comfort to the patient, but an adequate means 'of restoring **the** limb to nearly **a** straight position, though some degree of **false** anchylosis followed the extensive cicatrization of altered **or** destroyed parts. **The** turning **of the** cross-stick, to twist **or untwist the rope, was left to the wife of** the patient, who, under **his direction, relaxed or** tightened up the extension.

The **amount of movement of one bone** away from the other **is necessarily limited by the length of** the articular ligaments, which, **in their normal state, are adapted** to hold the bones in very **near** proximity. **If,** therefore, **the** attempt is made to regulate the extension by any measurements, there will be a strong temptation to employ **too much** force, producing a painful and unnecessary tension of **the ligaments** holding the bones **of** the joints in proximity. Where the limb has already become flexed, the amount of extension which would have been comfortable during the preceding inflammatory process may be safely exceeded, but while the limb has not yet lost its straight position, the extension should not be permanently practised to the extent of painfulness.

Barwell fully indorses this treatment in the confirmed stage of inflammation, though he fails to advise it in the very beginning. Quoting from p. 328, he says: "We have only to prevent the muscular spasm from pressing the two portions of bone together, and the disease will decrease. . . . . . . The muscular contraction which pulls the thigh up must be met by another which pulls it down. We cannot, nor do we wish to separate the bones, but we can so arrange that the muscular force shall expend itself upon an external object, and leave between the head of the thigh and the acetabulum no more than the normal amount of pressure, perhaps less. . . . . . These means will not cure the disease, but they will place it in the best possible circumstances for getting well."

DISTORTIONS AND STIFFENINGS AFTER THE SUBSIDENCE OF INFLAMMATION.—The distortions and stiffenings which result from cicatrization and anchylosis, after extensive destruction of the tissues of joints, become the most intractable cases. Restoration of motion and of form, by gradual extension, is discouraging, from the slow nutritive changes of the ligaments, the cicatricial tissues, and the "contractured" muscles; and the sudden rupture, by violent force, may excite an unwelcome inflammation, or be itself altogether impracticable without tearing open the integuments, or fracturing the bones, or both, which might involve dangerous nervous shock or subsequent constitutional sympathetic irritation.

The following illustration of a tibia partially dislocated backward upon the femur, and permanently flexed, from inflammation of the joint, and the disproportionate pull of the flexors which arises from irritation, affords a good conception of a class of irremediable deformities. Yet it would have been

simple and easy entirely to prevent this flexion and dislocation, by extension applied before the deformity commenced.
The extension would not only have prevented the deformity,
but it would have diminished the inflammation and the muscular contraction, by lessening or suspending the irritation
arising from the contact and frictional movements of the opposed joint surfaces.

Fig. 11.

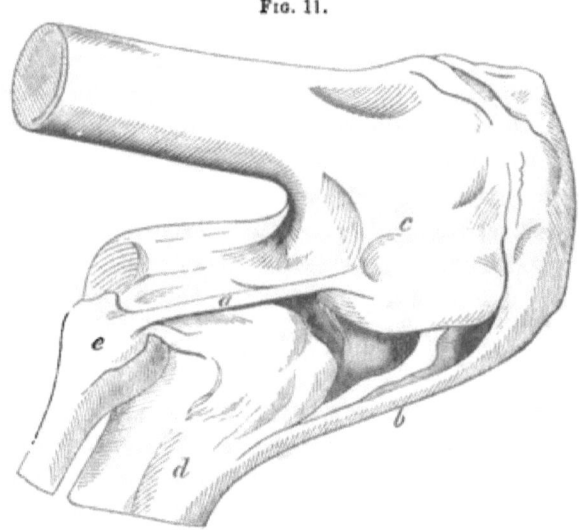

a, External lateral ligament.   b, Ligamentum patellæ.   c, Femur.   d, Tibia.   e, Fibula.

Figure 11, from Tamplin, illustrating the extreme flexion of
the knee amounting to dislocation, arising from the effort
of nature to bring the tibia into a position in which it cannot press upon the articular surfaces of the femur, in long-continued inflammation of the knee-joint, affords a hint at the
method of applying mechanism in order to correct the deformity.   It is obvious at a glance that any attempt directly to
extend the tibia upon the femur, can have no effect in causing
the head of the tibia to glide around upon the femur to restore
the proper rotundity of the knee and the natural play of the
articular surfaces.   If the effort is successful in bringing down
the tibia, its upper extremity must remain in such a malposi-

tion as seriously to interfere with the firmness of the joint
and its natural freedom of movement.

On the other hand, no attempt should be at first made to
change the angle of flexion.   The leg should be extended from
the femur in the direction in which the disease has left it, until
the muscles and ligaments become elongated.   The parts are
thus brought to a condition in which the extended position
may be gradually secured by the proper gliding motion of the
joint surfaces.

The neglect of this very important distinction has doubtless
led to many failures, when success was practicable.   It applies
to all extremely flexed joints, especially to the hip, to the
elbow, and to the shoulder; and it may be succinctly stated
as a rule, *to extend the tendinous and ligamentous fastenings
first and change the angle afterwards.*

It will be easily seen, that an extension in the direction of
the acquired deviation may be borne, which could not be tolerated if applied directly to correct the malposition, on account
of the extreme pressure of joint surfaces upon each other,
which must be produced by the long leverage of the bone attempted to be reduced to position.

This distinction has been most clearly
drawn by Dr. **H. G. Davis.**   The usual
appliances for extending limbs which
are permanently flexed with stiffened
joints, are illustrated by these plans
from Andrews, Figs. 12, 13.   One for
the hip-joint and the other for that of
the knee.

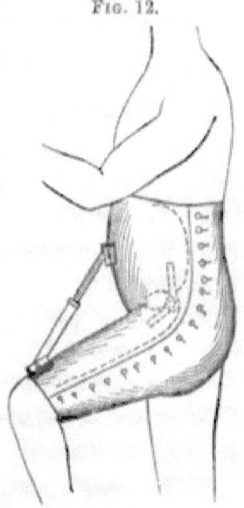

FIG. 12.

The artist has violated one of the
principles inculcated, by making the
brace in each figure, which is lengthened by a screw, apply itself to the
extremities of the splint.   Greater
power is obtained in this way; but in
any condition approaching that of Fig.
11, the head of one bone will tend to
revolve in its new position, instead of

Apparatus for reducing flexure at the hip-joint (from Andrews).

sliding to its proper place. Besides, the pressure fails to be equally distributed through the internode, as would be the case if the pressure were made upon the middle part of each portion of the double inclined splint.

Another objection to the figures is the want of elasticity, which should be secured by elastic rubber, or coiled wire.

FIG. 13.

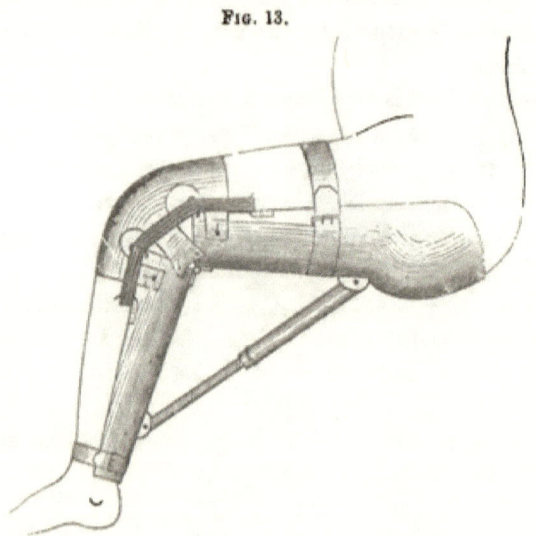

Apparatus for reducing flexure at the knee-joint (from Andrews).

By allowing the two parts of the brace, which extend from one portion of the splint to the other, to glide, one in a groove of the other, and attaching a strong rubber spring (accumulator of Barwell) to projecting pins upon the overlapping ends, the contraction of the rubber lengthens the brace, but not immovably.

The patient gets, by this means, some relief from the continuity of pressure, so that the circulation is better kept up, and ulceration avoided. A similar apparatus may be applied to the upper extremity for permanent flexion, with the same provision for elasticity.

There comes sometimes, an immobility of the radius upon the ulna, following fractures (without any maladjustment of the fragments), and also following rheumatism, in cases in which the deposit has not been fully absorbed.

The patient can be instructed to produce passive motion with the hand of the other side, and to employ machinery for the same purpose, which, from being more ceremonious, may come to be more faithfully employed.

Fig. 14.

This figure, from Bonnet, will afford a hint at methods of construction.

Sometimes, cases, apparently intractable, yield to mechanical force, proving abundantly successful. I quote in point, an instructive case, also from Tamplin, p. 169:

"In one case" (of false anchylosis), "which arose from a

cart-wheel passing over the thigh, just above the knee, there was motion, but of an elastic kind, and, after the leg had been forcibly extended, it returned with an impulse to its contracted position, without the flexors acting in the least, so that it was evident there was some adhesion in the joint. The flexors I divided, however, and the extension was kept up for some weeks, with but little benefit, and, from the severity of the increasing forcible extension necessary, the patient suffered so much that he left the institution, determined to submit to no more treatment. But, after a few weeks, he returned penitent, when I again operated, and resumed extension. We progressed, as heretofore, very slowly. From the constant pressure kept up on the anterior part of the thigh, a slough made its appearance, upon the healing of which I resolved to make no more attempts, and informed the patient that I feared nothing further could be done for him. This frightened him, and he set to work to screw his leg straight, at all hazards, upon doing which, a sudden and loud snap was heard, compared by the other patients to the report of a pistol, and immediately the leg could be placed in a straight position. . . . . . . He has left with a comparatively useful limb."

In the light of the present advancement of the art, the ligaments of this joint should have been first elongated by persistent extension within tolerable limits, and the limb straightened afterward.

The division of tendons, and such ligaments as can be reached, may greatly aid the effort at restoration, and, perhaps, if there is any degree of motion, the attempt should be made to convert useless limbs into useful ones, for, after the failure of strenuous efforts, the patient is apt to be more reconciled than before.

The division of the muscles themselves, may sometimes greatly aid in the restoration of the proper position. In the vicinity of large arteries, however, the subcutaneous division is not free from danger, however skilled the operator may be in anatomy, for their may be deviations from the usual relations.

It is probable that *over-extension* or interstitial rupture of *contractured* muscles, practised while the patient is in a state of complete anæsthesia, will become very generally substituted for tenotomy and myotomy, hitherto often found necessary in those cases in which the sarcolemma and fascia, among and around the fasciculi, resist the amount of extension which can be endured in the waking state without great suffering, and also in those cases in which a degree of extension which can be tolerated, increases the irritation and the permanent spasm from reflex action, aggravating the nocturnal starts and the whole train which follows irritation.

These operations of tenotomy and myotomy, performed to relieve the irritative contraction of muscles connected with the progress of joint-inflammation, are done subcutaneously, and in a state of complete anæsthesia, in order to get rid of the voluntary and reflex causes of contraction; and as there is always some danger of troublesome hemorrhage and subsequent suppuration, it may be well, while the patient is asleep, to practise the extreme extension, and if, after the expiration of a few days, the expected relief is not afforded, to divide the obstinately contracting muscles or their tendons.

In bony anchylosis, or in close false anchylosis, in which the ligamentous joining resembles that of an ununited fracture, the only remaining resource is Barton's operation, which consists in sawing the bone and removing a piece of such a shape as to permit the restoration of the limb nearly to its original general direction. This operation was originally practised by Dr. Barton, of Philadelphia, upon the lower portion of the femur, and has since been repeatedly performed.

This operation was performed upon the femur, above the trochanter minor, by Dr. L. A. Sayre, of New York, in 1862, making a movable and useful joint; a tolerable substitute for the hip-joint, which had become motionless from anchylosis. (See Transactions N. Y. State Medical Society, 1863, p. 103.)

IV. ACCIDENTS, BREAKING, TEARING, BRUISING, BURNING,
OR FREEZING THE TISSUES, FOLLOWED BY THE LOSS OF THE
PARTS, OR THE FAILURE, IMPERFECTION, OR PERVERSION
OF THEIR UNION OR RESTORATION.

The deformities resulting from breaking the bones and tearing the ligaments are more easily prevented than removed.
These accidents are sometimes of such a nature as to render
some degree of deformity unavoidable; but in those cases
much may be done by skilful management, to obviate permanent malpositions and ill-shapes.

The discussion of the treatment of fractures and dislocations
is outside of the scope of this paper. Some of the deformities
after fractures admit of great amendment by bending the new
formed bone, when the deviation of the long axis has occurred
from neglect of extension or lateral support, or from premature
use of the limb, and when the remedy has not been delayed
too long; and at a later period, by refracturing the bone and
keeping the fragments in proper relation to each other during
the period of repair; or by drilling the bone to secure inflammation and softening,* and, after about ten days, employing
force, applied gradually, to secure a change of nutrition with
the changing shape, or applied suddenly, to overcome the deviation by interstitial fracture, without complete solution of
continuity of the bone.

After severe bruises, with or without laceration of parts, the
destructive inflammation and gangrene, with extensive sloughing of intermuscular tissue, in the progress of areolar erysipelas, the greatest attention to passive movements may be
necessary during the period of repair, to prevent the formation of permanent adhesions and malpositions, and to stimulate
the absorption of the organized exudations, before they have

---

* The expedient of drilling the bone was performed by Dieffenbach for ununited
fracture, putting ivory pegs into the holes. Dr. Detmold, of New York, practised
the drilling without the pegs in 1850. Dr. D. Brainard, of Chicago, reported exper
iments with this expedient in 1854. See New York Medical Gazette, edited by Dr.
D. Meredith Reese, October 12, 1850, p. 232, and Trans. Am. Med. Assoc., 1854.

assumed such density of structure as to render them extremely difficult of elongation. In consequence of this, it may be necessary to divide them, in order to secure the proper restoration of position; and where the integuments are involved, to perform plastic operations, to cover the affected parts with healthy integument.

After freezing, the indication is to secure sufficient slowness in the thawing of the parts, by the employment of snow or cold water. The sloughing of integument may, in some cases, result in cicatrices, resembling those from burns, and requiring similar treatment, both for prevention of deformity and for its correction.

Burns result in some of the most frightful distortions, exceedingly difficult to prevent by extending appliances, and more difficult to remove by any means except the excision of the cicatrix. It is surprising to see into what a minute space a cicatrix will contract, when resistance to its contraction is removed.

The true indications in the treatment of burns, with reference to the prevention of deformities, are to moderate the intensity of the inflammation, first, by the local employment of turpentine or other capillary stimulants, and the internal administration of opiate and alcoholic remedies, to counteract shock and to annul or diminish reflex action; then to exclude air and other local irritants, by the application of flour, eggs, lead paint, poultices, &c., and to control the general circulation by alvine evacuants and veratrum viride, or other arterial sedative; and, finally, with the commencement of granulation and cicatrization, to support the general powers by nourishing food, and tonics if need be, and to counteract by mechanical appliances the drawing of the cicatrix. By this means, the shortening of cicatrices on the limbs will stretch the sound skin chiefly in the direction of the circumference; and when the contraction of the cicatrix has become complete by the yielding of the skin in this direction, the danger of distortion from the longitudinal pull is chiefly past.

To elongate a contracted cicatrix, in order to remove distortions already produced by its contraction, is altogether impossible. The correct treatment is, to remove the cicatrix and supply its place by sound skin taken from neighboring parts by a plastic operation; and, where this is impracticable, to make a free incision through the narrowest part of the contracted bridle, to straighten the limb and hold it in its restored position, until, by the contraction of the cicatrix, which forms anew, the sound skin has been drawn in from other directions. In some cases, an elastic retention will resist the pull of the cicatrix, at the same time that the functions of the muscles and joints are preserved, and the condensed sub-cicatricial tissue is more readily loosened up by the to and fro movement, and the surrounding integument is more speedily and more perfectly enlarged, to compensate for the contracting cicatrix. A rubber spring is most convenient for this purpose. Where the situation is not upon the limbs, but upon the neck, chest, or abdomen, the resistance to the contraction of the cicatrix is more difficult, and is sometimes impossible. Adhesive plasters, into the middle portions of each strip of which a portion of a rubber ribbon is inserted, promise more than any other expedients. The plaster must inevitably slide somewhat, and the rubber will secure an efficient pull, notwithstanding the sliding, and that, too, without so much interference with the ordinary movements as plaster alone imposes.

For closure of the mouth, generally resulting from *contracture* of the masseter and temporal muscles, or from the contraction of a cicatrix consequent upon ulceration or sloughing, a very convenient method of extemporizing a lever with which to act upon the jaws through the teeth, is to straighten out the handles of a tooth-forceps and put a thumb-screw through one of the handles.

For the purpose, however, of elastic extension, an instrument may be made on the principle of forceps, in which the tendency of the spring is to open instead of closing the blades. It will require very little ingenuity to extemporize an instrument with an elastic rubber spring, imitating those shears in which the blades open by closing the handles.

Among the deformities arising from perversion of nutrition short of inflammation, and in many cases not easy of **explanation**, are softening of **bones** in rickets in children, and what in adults we translate from good English into Latin, as the easiest means of getting a technical expression, *mollities ossium*, resulting in distortion, in obedience to preponderance of weight or muscular contraction, and ending **in** permanent distortion, as the lesion of nutrition disappears.

On the other hand, a hypertrophy occurs, producing an inequality of symmetrical parts and a deviation to one side **or** the other **of central organs.**

The **septum of the nose, sometimes, in** the period of **growth acquires a deviation to one side, diminishing or** closing **one nostril, and affecting** the appearance. **There** is no remedy **for this but steady** pressure, and the best fixed point is ob**tained by an** inflexible band around the head, from the front **part of** which an elastic rod may pass down by the side of the nose, where a suitable pad may be the intermedium of **the** pressure upon the nose. An extemporized band may be made by making a hat band of a piece of hoop iron attached to a hat, which has the shape of the head. The rim and crown **of** the hat having been cut **away,** a sleeping cap is the result, which can be laid aside during **the day. The** pressure upon **the septum will be more tolerable by inserting** a short piece **of a rubber tube into the nostril, on the convex side of the distorted septum,** which **of course, while the pressure is not applied, will tend** still **more to keep** the affected nostril open.

V. Mutilation of Parts designedly done, or Force or Restraint artificially applied, like the Compression of the Feet of the Female Children of Chinese Grandees, the Heads of some American Indian Tribes, and to a less degree the Feet and Waists of genteely educated Children in Modern American and European Society.

We laugh at the tastes and usages of people who attempt to secure distinction for themselves and their children, by compression and mutilation of various portions of the body, so that a lady has the **inimitable** distinction of requiring to be carried everywhere she goes, because her feet have been purposely made useless to her ; a chief, the high honor of having **a head shaped like** an idiot's, or the face slit in various parts for the insertion of trinkets. **But it** may justly be claimed, that these exhibitions of taste are reasonable, **compared with the** practice among **us of** compressing **and staying up the** chests of growing girls, so as to diminish the **capacity of the** lungs, and so to weaken the spinal muscles as greatly to **favor lateral curvature.**

Ambrose Paré, in the sixteenth century, appropriately remarked : "**I may not** omit the occasion of crookedness that **seldom** happens to country people, but is much incident to the inhabitants of towns and cities, which is by reason of straightness and narrowness of the garments that are worn by them, **which** is **occasioned by the folly of mothers, who,** while they **covet to** have **their young daughters' bodies so** small in the **middle as** may be possible, **pluck and draw their** bones away, **and** make them crooked."

A strange sentiment **prevails among us, that narrow** feet **indicate** refinement, **and broad feet,** vulgarity; **and so,** forsooth, **every** mother, **for her** children, **and all vain** youth of **both** sexes for **themselves, must compress** the feet with shoes too narrow for their **size;** until a portion of the toes ride **upon the others, and the** whole **foot is so** compressed as to fail **of the graceful** elasticity natural **to it.** Not the least deplorable **condition of the** deformities **resulting from these** usages, **is** the **impossibility of removing in the adult the** misshapes acquired in childhood **and youth. A deformity,** admitted to be such, may be **subjected to treatment early; but** a deformity, deemed fashionable, **will receive no attempt at** remedy, until **suffering** prompts the victim, **too late, to seek relief.**

There is **one possible good of these deformities;** that is, to convince ignorant people that, as **the body is** capable, in the **growing** period, of having these **abnormal** shapes purposely produced, so, by appliances purposely made, deformities may **be gradually corrected.**

PART SECOND

PARTICULAR DISEASES AND DEFORMITIES,

Not yet Noticed, or only Incidentally Referred to.

INFLAMMATION OF THE JOINTS OF THE EXTREMITIES AND RESULTING DEFORMITIES.

I. HIP DISEASE—MORBUS COXARIUS.—The circumstances of inflammation of this joint single it out as the lion of its class. It is the joint to which more force is applied than to any other joint in the body, with more varied and more extensive movements, with a closer contact of surfaces, and affording more extensive friction surfaces when the parts are inflamed. It is surrounded by stronger and more numerous muscles than any other joint, which, when spasmodically contracting, subject the head of the femur and the upper surface of the acetabulum to great pressure. On this account, synovial and cartilaginous inflammation, when once initiated, is more aggravated by movement and pressure, than in any other joint. The depth of the joint may be the reason for the obscurity of the sensations of discomfort in the early period of the disease, and for their reference to the parts lower down, and especially to the knee-joint.

The diagnosis of inflammation of the synovial membrane and articular cartilages of the hip-joint, commencing in the usual subacute form, is so obscure as never to be made out by a non-medical observer in the early period.

The patient, if a child, is observed to trip and fall down with unusual frequency, and to complain, especially at night, of pain in or about the knee-joint.

This leads parents to suppose that the disease is in the knee, though there is neither swelling nor sensitiveness to pressure in or about the knee. No movement of the limb in the early period will give rise to pain in the hip-joint, because the parts thus far affected are destitute of nerves of sensation.

It may be observed, however, that there is a flattening of the gluteal muscles of the affected side, compared with the opposite, and that in walking, the movements of the affected side are less free than those of the opposite. Pain in the hip-joint itself, is a later symptom, arising after the sensitive nerves of the bone have become involved, and accompanying an apparent shortening of the limb.

An apparent elongation of the limb is an early symptom, attracting the attention of parents and nurses before there is usually much complaint by the patient.

This apparent elongation of the limb arises from the tilting of the pelvis, and may be a movement, in which the lumbar and ilio-lumbar muscles relax, to secure sympathetic relaxation of the ilio-femoral muscles of the same side, and thus diminish the pressure of the joint surfaces. After the period of muscular spasm has arrived, this is all reversed; the lumbar and ilio-lumbar muscles contract in their turn, from sympathy with the irritant contraction of the ilio-femoral muscles.

This primary dropping of the affected limb may be taken as a hint at the correct mechanical treatment in all stages. This branch of treatment, in the early period of the disease, has been sufficiently discussed; but in the periods of more aggravated suffering, its importance needs still more to be enforced; and this can best be done in connection with the farther notice to be taken of the pathology. As a most impressive presentation of the subject, attention is invited to the following quotations from Barwell " On the Joints," p. 312 :

" The nervo-muscular phenomena in hip-disease are so prominent and remarkable, that their evident results, as seen in the posture and apparent length of the limb, have chiefly attracted the attention of surgeons; and yet, the peculiar influence which they have upon the continuance of the malady, has escaped notice. Be it observed, that the constant and violent contraction does not simply produce ad- or ab-duction, according as one or the other set of actions may prevail; but as, from the direction of the muscles, it is evident, it must draw the thigh up and cause the head of the femur to press abnormally against the acetabulum. Thus, the pristine inflammation, having produced a contraction, the head of the thigh-bone begins to press with abnormal force and constancy in the upward direction. . . . . . . To prove this position, we have only to look at a pathological museum. We shall find a few specimens in which the action is distributed over the whole joint surface; a very few indeed, in which the inflammation has chiefly attacked the lower portion or anterior part of the acetabulum and femur; but in a proportion so large as to render the above examples mere exceptions, the upper lip of the cotyloid cavity and the corresponding portion of the head of the femur are ulcerating, while all the rest of the bone may be untouched. Such constancy of action can only be accounted for by the fact that abnormal muscular contraction produces pressure, and thereby ulceration of the parts. . . . . . Thus, the acetabulum is made to travel upward and also inward, whereby an opening into the cavity of the pelvis is not unfrequently produced. I say, that such evident yielding to pressure is not an exceptional case, but is the rule; that, when we find a hip-joint ulcerating in any other way and position, it is that some rare circumstance has caused a primary ostitis in that particular spot. It must also be remarked, that as the head of the femur travels upward, producing ulcerative absorption in that part, against which it so abnormally presses, it causes beyond that point, an additional growth of bone, forming a new lip to the new cavity."

Page 317. "*Dislocation* of the head of the femur, from disease, or spontaneous dislocation as it is called, is an occurrence so unusual, that one is astonished at the general credence in its frequency. It is only about ten or fifteen years ago, that every hip-joint disease was supposed to end in this way; but, if a search be made in the College of Surgeons, St. Thomas's, St. Bartholomew's, or others of our great pathological museums, there will be found very few specimens exhibiting, simultaneously, the signs of morbus coxarius and spontaneous dislocation.

"On the other hand, it is by no means uncommon, to find the head and neck of the femur shrivelled to little more than a button-like projection, the acetabulum quite altered in form and place, and yet the bone retained in its cavity. . . . . . . Spontaneous dislocation occurs only in cases of so cachectic a character that new bone is not produced beyond the focus of suppuration."

In the Wistar and Horner Museum, University of Pennsylvania, Philadelphia, is an interesting specimen, No. 2148, of destruction of the greater portion of the head of the left femur, in an adult. The surface of the acetabulum is roughened, while its rim is elevated by new growth of bone making a deeper cup. The surrounding bony surface is roughened, but there is no evidence of there having been dislocation.

A rare and interesting specimen of bony anchylosis of the hip-joint, is in Dr. Pancoast's collection in the Jefferson Medical College, Philadelpha. The specimen is that of an adult, but is not numbered.

The head of the left femur is fused upon the lower rim of the acetabulum with partial dislocation directly downward, so as to leave empty the upper half of the acetabulum, which at the same time is partially filled with bony deposit.

The head of the femur is at the same time rotated, so as to carry the shaft of the femur directly up along the side of the trunk.

Specimen 215, in the cabinet of the Medical College, Boston, exhibits an anchylosed hip-joint, in an adult.

The trochanter major lies so closely upon the ilium, as to show that the head of the femur had been in great part destroyed, before the joint became obliterated by the new bony deposit.

This case was evidently one of *morbus coxarius*, and not one of rheumatism.

Apparent Dislocation.—These figures (Fig. 15), afford a good illustration of what passes for dislocation after hip dis-

Fig. 15.

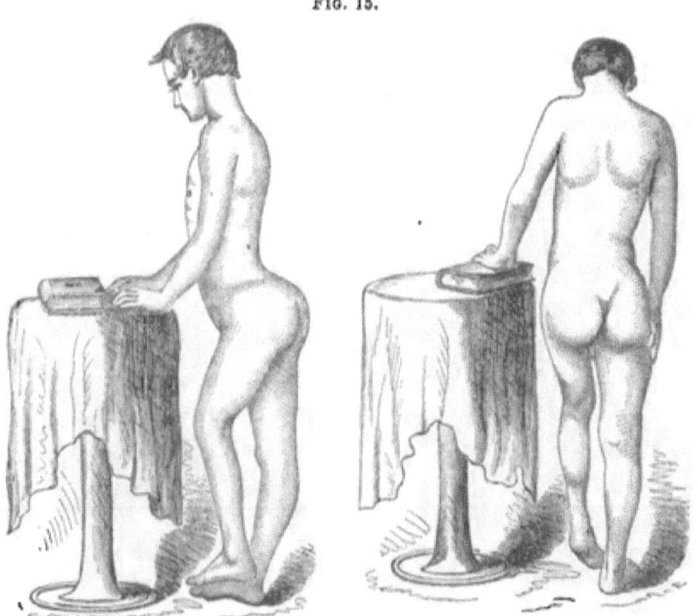

Apparent Dislocation after Hip Disease.

ease. The half-grown boy, from whom these sketches were taken, had recovered from hip disease without mechanical treatment, and his physician, an intelligent man, supposed the shortening to result from dislocation. It was soon shown by measurements, that the distance from the superior anterior spinous process of the ilium to the internal malleolus, was as great on one side as the other, when the sound side was brought forward to correspond with the other. The apparent shortening was made in part by a lateral tilting of the pelvis, and in part by the permanent flexion of the thigh upon the pelvis.

The case was also supposed to be one of anchylosis, but a careful examination proved the existence of motion, though only to a very limited extent.

The case is introduced as a representative illustration of the sequel of scrofulous inflammation of the hip-joint, proceeding to destruction of substance of ligament and bone, and recovering without mechanical restraint to obviate the stiffness and permanent flexion with adduction and inversion which naturally result.

The illustration will also help us farther along to appreciate the compensating capability of the spinal column.

II. INFLAMMATION OF THE KNEE-JOINT appears in the same classes of tissues, in the same forms, with the same course and termination as in the hip-joint, only that the thinnness of covering over the knee-joint renders the early diagnosis less obscure. There is less danger of being misled by the sympathetic pains which so often deceive in the early period of hip-disease. The indications for treatment are in all respects the same, and the facility of extension is greater. While the swelling, and tenderness on pressure, about the hip-joint sometimes renders it necessary in treatment to make counter-extension from the chest, this expedient is rarely necessary in disease of the knee-joint. The cases of *white swelling* of the knee, in which chest counter-extension will be most often found advantageous, are those in which extreme flexion has already occurred, with purulent burrowing tracks among the muscles of the thigh. By this mode of applying extension, the posterior part of the leg will be the only part of the affected limb subjected to pressure. This mode is equally applicable to flexion of the knee and of the hip-joints, and permits a considerable freedom of movement to the patient. With elastic fastenings, the muscles are permitted to contract and relax, securing to the joints a healthier nutrition than attends a uniform unyielding pull. At length, when the limb is brought out to some approach to a straight position, the counter-extension can be changed from the chest to the groin or the ischium, or to both combined.

III. INFLAMMATION OF OTHER JOINTS.—Inflammation of the ankle, of the tarsal and phalangeal joints, hardly admit of **extension,** nor is deformity one of the frequent results, though stiffness may ensue. In the upper extremity, **extension is** practicable, but not so necessary, because the muscles are less strong, and in walking, there is no push upon the bones, as in the lower extremities, and besides, in the upright position, the weight of the limb amounts to a moderate extension.

Inflammation of a metacarpo-phalangeal joint may require extension, to avoid both stiffness and deformity, and this **is** easily effected by passing a rubber ribbon over the space be-**tween the thumb and the forefinger, and attaching** each end of this **ribbon to the proximal end of a small splint, reaching from the lower end of the radius, to a point a little beyond the end of the affected finger, to which** the distal end **of the splint is attached by isinglass** plaster. The slight but constant **pull, may save the** finger **from** being rendered useless by stiff-**ness or** permanent flexion, or both combined, at **the** same time that the extension is **a** source of comfort to the **patient.**

Apparatus for Extension.—The pioneer in **the introduction of** practicable forms of apparatus to answer the **purpose of ex-**tension, is doubtless, Dr. Henry **G.** Davis, of **New York. In a** paper which he published **in the American** Medical Monthly, for May, 1856, referred **to in his paper in the** Transactions of the American **Medical Association,** 1863, he **says: "** There is one point **in my mode of making extension, which I think, from** the long experience I have had in its use, would be an improve-ment in the general mode, and it is equally applicable to all ex-tensions and counter extensions, those of fracture as well as contracted muscles, viz., the use of rubber as an extending power. This will act steadily and gradually, without any vio-**lence, and with very little** suffering, in comparison with perma-nent fixtures. When contracted muscle is to be overcome, it steadily wearies it, until it quietly comes off conqueror."

We are permitted to copy the following cuts and descriptions of apparatus from Dr. Davis's report in Trans. Am. Med. As-**sociation,** for 1863:

"The apparatus, by means of which my treatment embodying the principles advocated in this paper can be carried into effect, is simple and easily explained.

"It must be borne in mind, that I have already said that the essential parts of the apparatus are, means of exerting an elastic, continually-extending force on one side, and a resisting counter-extending one on the other.

"The modifications it undergoes to adapt it to the various regions of the body, every physician can readily understand. I shall describe particularly the splint as applicable to the hip-joint. Reference to the wood-cut (Fig. 16) annexed will further aid the reader.

Fig. 16.

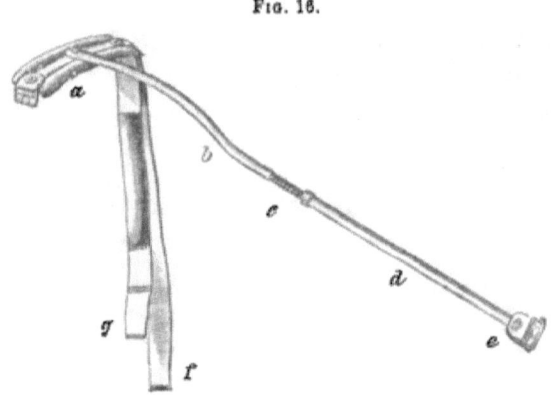

"An elastic perineal band (g in the figure) really constitutes the extending, adhesive plaster strapping around the limb, concentering at a point a little above the external malleolus, the counter-extending power, while a metallic splint (b, c, d, e), is stretched between these, and enables them to fulfil the indications proposed.

"The splint is composed of four parts, viz., an upper or pelvic portion (a in the figure), a thigh portion (b), a leg portion (d), and an ankle portion (e).

"The thigh portion consists of a metallic tube curved above (b), to correspond to the convexity of the thigh, and ending below in a short, straight piece, to which a long iron double-threaded screw (c), also hollow, is firmly secured. The leg portion is a straight metallic tube (d), closely investing the screw (c), and projecting but little beyond. It is attached to the ankle portion (e) in such a way that, in revolving it, in order to lengthen or shorten the instrument, only the screw investment with its nut is turned, while the ankle portion remains unmoved. This ankle portion (e) consists of a triangular-shaped, flat piece of metal, covered by a buckle (as seen in the figure).

"The pelvic portion (a) is more complicated than the others. A slightly curved strip of steel half an inch or more wide, from four to six inches (both the length and width varying according to the size of the whole splint) long, is riveted through its centre to the free extremity of the thigh portion, and admits of rotary motion. At one end of this strip of steel a buckle, and at the other the perineal band is attached, and the whole of it is well cushioned. The perineal band is formed of two bands (f and g), of a length, width, and strength varying according to the size of the apparatus and the circumstances of its application. One (f) is longer than the other and inelastic, being made entirely of strong cotton or linen webbing, the other (g) is, as it were, an oblong bag of India-rubber webbing (formed by sewing two strips of rubber webbing together) filled with sawdust, obtained by sawing across the fibre of pine wood (not lengthwise), tipped at each end with some of the inelastic webbing (such as (f) is made of). While the inside elastic band keeps up the extension required, the inelastic sustains any weight that exceeds the extending force as then applied to the patient. It is this arrangement that enables the weight of the body to be borne without harm, as in walking, and that prevents injury from excessive weight or pressure upon the articulating surfaces in cases of accident. Thus, for instance, the head of the femur would, in walking, be violently thrust upward, as the elastic band would yield to an increased weight, were there no inelastic, unyielding band to prevent it; yet, it is obvious that this inelastic band does not interfere with the predetermined amount of tension to be exerted by the elastic one. (This amount of extension is determined and regulated as follows: Buckle the two bands unequally, *i. e.*, let the loop formed by the outside band be longer than that of the inside, and attach a weight to the latter. The number of pounds requisite to stretch the one loop to the exact length of the other represents the amount of extending force the instrument will exert, when exactly thus buckled, when applied upon the limb.

"I will add that here the amount of extending force should be ascertained in every instance before fastening the splint upon the patient; this amount is not to be varied by altering, by means of the screw, the length of the instrument, but by adjusting the two bands.)

Fig. 18.

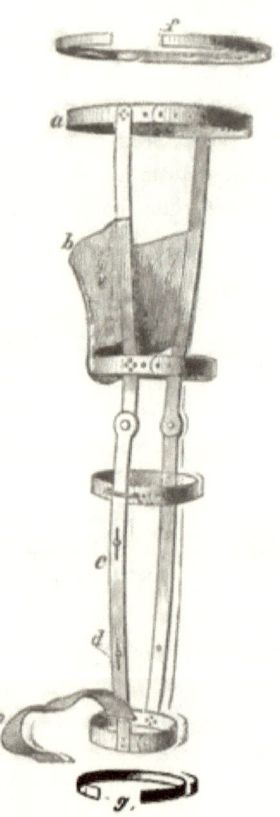

"I have foun¹ the rubber webbing, with napkin protection to the skin, to be much superior to the rubber tubing. The tubing is seldom of the right elasticity, and is more liable to heat and excoriate the skin than the webbing, for the reason that it does not absorb the secretions,

Fig. 17.

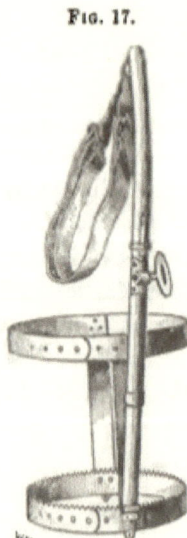

nor allow the air to come in contact with the skin.

"The granular stuffing moves easily on itself and displaces

readily, so as to equalize the pressure, as, for instance, over the adductor tendons in the groin.

" The long splint is best adapted to the majority of cases. Some years ago, I was in the habit of applying a shorter one (Fig. 17) to the femur alone. This leaves the knee at liberty, and in so far is an accommodation to the patient, but other-. wise is not so effectual.

" I also generally use the long splint for disease of the knee-joint, applying the adhesive plaster only to the tibial portion of the limb. I have devised, however, a very convenient instrument (Figs. 18 and 19) for extension at the knee-joint, that admits of flexion and extension of that joint, which in cases not too severe is sufficiently effectual.

"Extension should be constant;

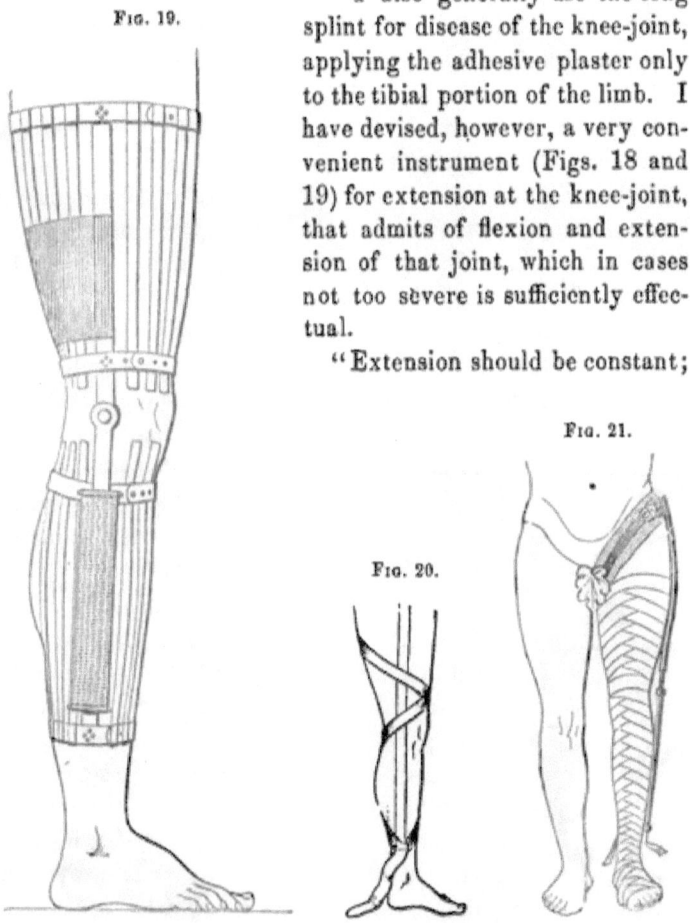

FIG. 19.

FIG. 20.

FIG. 21.

when not accomplished by the splint, it should be by means of a weight and pulley.

"*Mode of Application.*—Cut from a piece of adhesive plaster, spread on twilled goods and kept until the oil entering into its composition has become oxidized, two strips from 1¼ to 1½ inches wide, of the length of the limb from the pubis to the malleolus, and two strips a little narrower in proportion to the others, but one and a half times as long. Fold about an inch and a half of one extremity of each of the first cut strips upon itself, the adherent sides to each other, and apply one on the outside and one on the inside of the limb, commencing with the folded end about two inches above the outer and inner malleoli, and extending it up in a straight line.

"The other two strips are applied spirally around the limb as follows: Commence on the lower or folded extremity of the straight strip above the outer malleolus, and wind around in front and back, so that the two spiral strips meet in front, a little distance above the patella (as very well depicted in the wood-cut Fig. 20).

"Next, sew a piece of firm, *inelastic* (linen or cotton) webbing, about 1¼ inches wide and six to eight inches long, to the lower extremity of each straight strip, taking particular

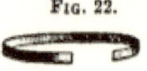

FIG. 22.

care to include in the attachment the ends of both spiral strips above the external malleolus. The limb is then closely and firmly enveloped with a common roller bandage, from the foot upwards (as shown in Fig. 21), the pieces of webbing only being left outside free. Now buckle the ankle portion of the splint upon the external face of the limb by means of the webbing; protect the skin of the groin and parts to be covered by the perineal band by a piece of old, soft napkin or table linen, several times folded and secured by a few stitches; and having previously adjusted the two bands composing the perineal band, as already mentioned, fasten the latter around the thigh, always taking care to have the buckle on the pelvic portion of the splint in front; the

screw of the splint regulates its length, so that the
required amount of extension can be secured. When
all is correctly arranged, and proper extension

Fig. 24.

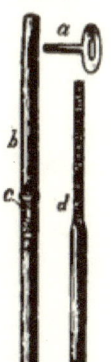

Fig. 23.

made, the upper extremity of the splint should fall just below
the crest of the ilium."

Figs. 22 and 23 are further modifications of extending apparatus, and Fig. 24 illustrates the manner of connecting and
fastening the two portions of the extending shaft.

Fig. 25.　　　　　　　　　Fig. 26.

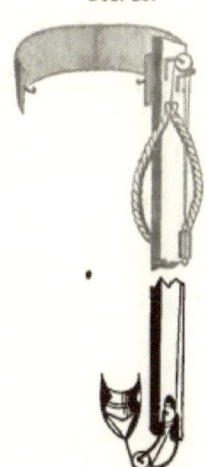

Dr. Sayre's splint, with a wheel and ratchet
extension.

Barwell's splint for hip-disease, with a
pelvic band for fixing the upper end of
the splint.

Dr. Vedder, of New York (now deceased), modified Dr.
Davis's splint, chiefly by making the extending shaft, in great

part, of wood, which slides in a metallic trough, and fastens
by a catch falling between the teeth of a ratchet. The counter-extension, as in the apparatus of Davis, is the usual
oblique band over the groin and perineum.

Dr. Lewis A. Sayre, of New York, has made a very compact
and beautiful modification of Davis's splint, shown in Fig. 25,
manufactured by Otto & Reynders. Provision is made for
adhesive plaster extension on both sides of the leg, by a bow
which makes a half circle anteriorly. In the figure, two
transverse pieces are shown for greater strength.

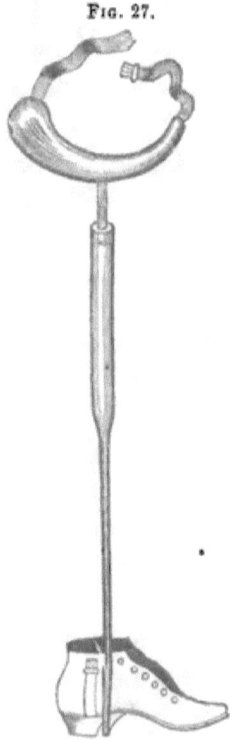

Fig. 27.

Dr. C. F. Taylor, of New York, has
also a modification of Davis's splint.
The cross-piece at the upper end of the
splint, with either a vibratory or a universal motion, seen in the figure of
Sayre's splint, which is not in Davis's
earlier apparatus, was suggested by Dr.
Taylor. The object of this is to prevent
the too close pressure of the perineal
band anteriorly, and to increase the ease
with which the apparatus fits in flexion
and extension of the thigh.

Mr. Richard Barwell, of London,
has also a less ingenious extending
apparatus, shown in Fig. 26.

Dr. Edmund Andrews, of Chicago,
has devised an extending apparatus, in
which the extending shaft is made of
gas-pipe, with a screw extension, as in
the apparatus of Davis; but Dr. Andrews places the splint upon the inside
of the limb, with a sort of crutch-head,
fitted to the shape of the coverings of

Dr Andrews's inside splint,
with ischiatic crutch-head, for
hip-disease.

the pubis and ischium, to receive the
weight of the body in walking, and to

be, at all times, the medium of counter-extension, or, more
properly, counter-pressure.

. All these forms of extending apparatus require more skill in constructing them than the general practitioner can ordinarily find in his nearest blacksmith and carpenter, and I have endeavored to perfect a plan of apparatus which will enable all practitioners, both in town and country, to give their poorest patients benefits of treatment which, for practical results, may be equal to that of experts in this particular branch.

After trials of various forms, I have settled down upon the plan illustrated in Fig. 28. The shaft is placed upon the outside of the limb, and a collar is adapted to the thigh, to press like a crutch upon the ischium.

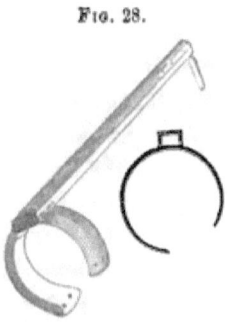

Fig. 28.

The whole splint (adapted to the left side) is seen in an oblique aspect. The shaft is of hard wood, and has the collar for the ischium upon the upper end (lower, as seen in the cut), and upon the other end, an iron foot-rod, which is intended to pass through the heel of the shoe. The shaft is made so long that the weight of the body cannot come upon the foot. The collar, which can be made of hoop-iron by any blacksmith, is about $\frac{13}{16}$ of a circle—$\frac{8}{16}$ behind, and $\frac{5}{16}$ before—and should be carefully padded.

Ischiatic crutch for extension in diseases of the hip and knee-joints. The small figure is a profile of the collar attached to the end of the staff.

For children, this should be temporary, and made of cotton and muslin, in order that it can be easily and cheaply renewed as often as it becomes wet.

The small figure is an end or profile view of the iron collar attached to the end of the wooden shaft.

The shaft may be made extensible and adjustable, but this requires the skill of a professional instrument maker, and is expensive, while, if made for a particular individual, there is no advantage in it.

In an extemporized apparatus, the ischiatic collar can be screwed on to the shaft, higher or lower, until the right point is obtained, when the shaft can be cut off above the collar.

The collar should be bevelled behind, to apply to the ischium without discomfort, but in front this is not necessary. The open space comes opposite the great vessels descending from the groin. The collar is such a segment of a circle as not to need a strap to hold it in position.

It frequently occurs that the swelling and sensitiveness of parts in the vicinity of the hip-joint forbid the use of the collar. The use of the perineal band is often forbidden, also, by the same condition, driving the operator to the employment of a weight over a pulley for extension. In this case, the weight of the body, while the limb is elevated, makes the counter-extension.

In many cases, counter-extension from the chest will be found more convenient, permitting the patient to be turned over, though subjected to the straight position.

With the disappearance of the local tenderness and swelling, more portable apparatus can be employed.

Attachment for Extension.—The readiest and most efficient expedient for extension, at the same time that no inconvenient projection is necessarily made beyond the foot, is that invented by Barwell, and figured in his little book, The Cure of Club-Foot. It consists in attaching a parallelogram of tin to the leg, by means of adhesive plaster, so as to get the point of attachment of the extending agent near the knee-joint. A spring of elastic rubber or a spiral steel spring can then be attached, above, to the upper end of this piece of tin, and below, to the lower end of the extending shaft, which may terminate either above or below the foot. Fig. 29 illustrates the mode of applying the tin and plaster.

For the purpose here contemplated, one parallelogram of tin should be placed on one side, and another on the opposite side.

By means of this expedient and the ischiatic crutch, extension upon the hip-joint and the knee-joint can be kept up while the patient walks about, the whole weight of the body coming upon the shaft of the crutch, and being received by the ischium. The muscles passing the hip-joint, direct the movements of the limb, but impose no pressure upon the hip-joint, their own contraction being counteracted by the spring which makes the extension.

It is convenient to attach a pretty long heel to the shoe,
with a hole in it just large enough to receive the angular spike

Fig. 29.

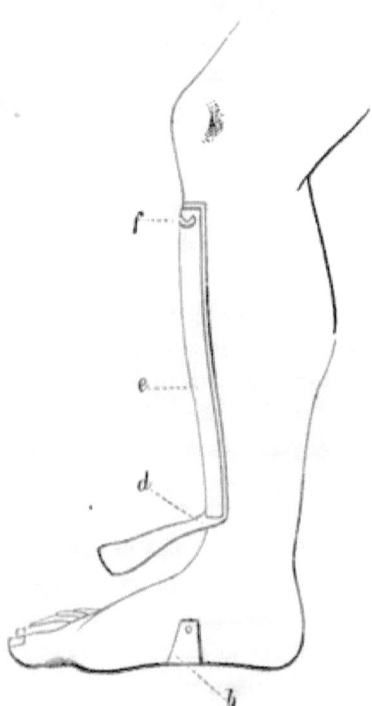

Barwell's method of attaching a piece of tin to the leg.

d The lower end of a strip of plaster attached to the leg, and upon the outside of which, the
tin, *e*, is applied. The plaster is then turned up against the tin, and circular strips of plaster
are applied covered by a roller. *f.* A loop for the attachment of the extending spring. *b.*
represents a loop of plaster attached to the foot in the treatment of talipes valgus. This is the
position which the termination of the splint occupies.

upon the lower end of the shaft.   As this passes through the
heel of the shoe, any pull upon the shoe expends its force upon
the shaft, through which it finally comes upon the ischium.
When the pressure upon the ischium becomes in any degree
uncomfortable, the patient puts his hand upon the upper end
of the shaft and pushes it down, and relieves the ischium from
all pressure whatever. As the rubber springs yield, the extension is increased, at the same time that the pressure upon the
ischium is relieved.

At night, the shoe can be taken off, and the elastic extending straps can then be attached to the angular spike, in order to keep up the extension while the patient sleeps.

By employing Barwell's method of applying the plaster to the thigh, the necessity of carrying the apparatus below the knee, may be avoided in some cases of hip-disease, whether the counter-extension adopted is that of the ischiatic crutch, or the oblique perineal band.

LATERAL CURVATURES OF THE SPINE—SKOLIOSIS.—Deviations of the spinal column to the right or left of the straight median line, or in one part to one side and in another part to the other side, must depend upon deficiency in the bones, the ligaments and cartilages, the muscles, or in some necessity of the spine to deviate from the straight line to tilt the pelvis toward a short or distorted limb, or to conform to changes in the volume of the organs within the chest, or upon several of these causes combined.

Figure 30 (from Bonnet), affords a good illustration of lateral curvature in which deviation in one part is compensated by an equal deviation in the opposite direction in another part.

Though such a perfect compensation of curvatures is not often met with, the figure shows very beautifully the natural tendency. The figure shows, also, a rotation of the bodies of the vertebræ in the lumbar region, by which they acquire a greater curvature than that of the spinous processes.

There are several specimens in Dr. Pancoast's cabinet, in Jefferson Medical College, Philadelphia, which illustrate this tendency. One of these is a compound of vertical and lateral curvature, the eighth, ninth, and tenth dorsal vertebræ, and also the second and third lumbar having become fused by new bone. Though the amount of material loss was very great, the reparative processes were commenced before the destruction became extreme.

A greater deviation is thus permitted than the actual elongation of ligamentous connections accounts for.

This element of twist in the mechanism of the curvature,

has been long recognized, but particular attention has been recently called to the subject by Barwell, in his work on the Joints, and by Sir W. Adams, in his work on Lateral Curvature, &c., London, 1865. The latter author goes to the extreme of accounting for the curvature by this twist entirely, and refusing to the assumed elongation of ligaments any share in the process.

The analogy of other articulations, however, gives us no reason to suppose that the ligaments of the spine should be exempt from the fate of those of other joints, though the extent to which the dorsal vertebræ are braced by the ribs, affords very considerable protection to the ligaments of this region.

The effect of the twist here, is to increase the curvature of the ribs at the angle on one side and to diminish it on the other. It is this change in the curvature of the ribs that constitutes one of the greatest obstacles to the restoration of the proper form, in confirmed cases.

The classification of these curvatures, in accordance with their pathological causes, must often be very much in doubt, because the opportunities for

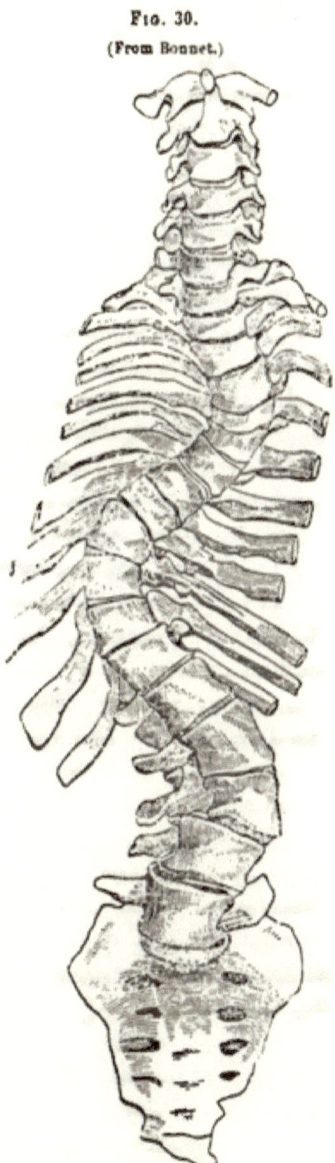

Fig. 30.
(From Bonnet.)

post-mortem examinations in persons dying from other diseases, during the early periods of lateral curvature, must be rare, and, when they do occur, the lesions may be so difficult to distinguish, or the primary alterations in one set of organs, and the secondary changes in another set of organs, may be so equal, that the distinction of cause from effect may be impossible.

In the present state of the science, the classification must be made more upon general considerations, than upon observed pathological states of the organs.

The simplest division is based upon,—

I. Supposed weakness of bones in proportion to the muscles rapidly developed.

II. Disease of the ligaments and bones, weakening them, not only relatively, but absolutely.

III. Fatigue of overtaxed, feeble muscles, shirking their work upon the ligaments.

IV. Spasmodic contraction of the muscles of one side, affecting a single muscle, or a greater or less number, from irritation in the brain, the spinal cord, or in the course of the nerves, or reflected from some place of irritation outside of the central nervous system.

1. A person, during the growing period, has a course of fever, and upon recovering, gains two or three inches in stature in a very short time, and acquires a curvature while the muscles are in active and healthful exercise.

This picture is in obvious analogy with cases of curvature of the bones of the lower extremities of infants (from some tardiness of development of bone, or from actual disease, as in rickets), in whom the muscles develop more rapidly than the bones, or in whom the ambition of parents or nurses leads to the too early teaching of the child to stand and walk. If the tibia were divided into segments, with intervening cartilaginous and ligamentous material, like the arrangement of the spine, it would, doubtless, bend; the bony substances, yielding as readily in the multiplied pieces as in the single piece.

The muscles, outrunning the bones in development in both

cases, the principles of treatment must be the same in both. The chief indication is, to place less weight, or for shorter periods, upon the bones.

. For cases of this class, sufficient rest in the horizontal posture is more important to be enjoined than any system of movements.

The movements which are practised, however, should be so diversified as to bend the spine in every direction of which it is capable, and so amusing as to be a pastime to the patient. Of all practices, however, sitting on a bench at school, is most injurious to a person of this class. Permanent distortions, from this class of causes, may require apparatus, as a bow-leg requires a splint, with persistent force to straighten it.

2. A young person, in falling from a height, or from some other form of violence, sprains the ligamentous fastenings of the spine, and after several months, a curvature, chiefly on one side of the median line, is noticed, of such a marked character as to seem to have occurred suddenly, while very little complaint is heard from the patient, whose lassitude, and indisposition to the sports of youth, are out of proportion to any feelings of pain or discomfort arising from the disease.

This picture is in plain analogy with those cases of white swelling of other joints, which originate in the synovial, ligamentous, and cartilaginous investments of the bones, and which are very slightly painful, on account of the paucity or absence of nerves of sensation in the tissues affected, or of their failure to be awakened into painful activity by the low grade of the existing inflammation ; producing changes of volume so slowly as to give the nervous filaments time to accommodate themselves to the changing relations.

It may be assumed that, when tissues affected with acute or subacute inflammation pass into a state of chronic inflammation at the ordinary period of the termination of the acute disease, there must be some continued causes of irritation, or some cachexia, original or acquired, temporary or permanent, which secures the continuance of the disease in the chronic form. In this sense, there may be a constitutional disease, requiring constitutional treatment, as well as rest, of the parts affected. The cases in this class differ from those of the first class, in which there is no disease ; only disproportionate development.

A very remarkable specimen of this class may be seen in an adult skeleton in the cabinet of the Medical College, Boston, No. 420. In this instance, the change of form is confined chiefly to three vertebræ; the eighth dorsal, which is wedge-shaped, and the two adjoining vertebræ, each of which exhibits loss of substance on the side corresponding with the middle of the three.

The bodies of the vertebræ exhibit no appearance of having been ulcerated. The absorption of the bony substance seems to have been produced by disproportionate pressure.

The claim of Sir W. Adams, that a twist of the spine is essential to lateral curvature, is completely disproved by this specimen, in which there is no twist, though the portions of the spine above and below the curvature make with each other an angle of about sixty degrees.

The confirmed cases of this class are entirely incurable.

As the first indication for treatment in all cases in which the disease is progressing, it is obvious that relief of the inflamed ligaments from the strain of the weight and movements of the body, is of the first and highest importance.

Nothing can do this so effectually as the horizontal posture; but if the deformity has become confirmed, and resists the attempt to strengthen the spine by the hands of the surgeon, the curvature will never straighten itself. Force from without must be persistently applied by suitable apparatus.

3. A growing person, generally a girl, of lax joints and slender muscles, is restrained from the diversified movements which are so delightful to children and youth, while the desire of securing a straight spine and slim waist leads to the application of stays, which interfere with the free play and rocking of the vertebræ upon each other, at the same time that, from the partial disuse of the lateral muscles of the spine, they acquire a more marked degree of atrophy than the other muscles of the body; while yet a monotonous sitting posture, under restraint

at school, presents the strongest temptation to rest the muscles, by permitting the weight of the head and shoulders to curve the spine, as far as the embracing stays and the ligamentous connections permit, which, by habit, comes at length to be always on one side in one part of the spine, and on the other side in another part.

This picture describes the class of muscular curvatures of artificial production, caused or aggravated by the means employed to prevent or cure them.

A girl left to choose her own amusements and occupations never would acquire this kind of curvature. The instinct of movement would lead to diversified action of the muscles, and, on becoming fatigued, she would lie down. Rest, in the horizontal posture, would be always more grateful and more complete than that rest of the muscles which is secured by allowing the weight of the parts to come upon the ligaments, and the surrounding artificial supports.

As far as a person can bend the spine, by a moderate effort, so far it will deviate in this kind of resting; and, coming to be habitual in one direction for each portion of the spine, the ligaments will at length elongate on one side, and shorten on the other, as in the production of deformity of other joints, until a habitual and fixed curvature is the result.

This class differs from the other two, in being the result of the vicious tastes and usages of fashionable society, or of the excessive regard for intellectual training, in disregard of the necessities of the physical constitution.

I am constrained to quote, on this subject, Dr. Beal, to show how well this was understood, thirty or forty years ago:

" Inactivity is not alone sufficient to account for that degree of muscular debility which induces spinal deformity. In warm climates, women take less exercise than they do in this country (England), and yet, curvature of the spine is infinitely more rare in such countries than in our own. In hot climates, all people indulge more in the recumbent position ; reposing themselves, when fatigued, as nature dictates. With us, a girl, from the age of ten, is obliged, throughout the day, to main-

tain a constrained position of the body. She is not permitted
to rest the muscles of the back, however weary, and the admo-
nitions of parents and tutors are unceasing, to keep herself
erect. In this way, the muscles of the back are overstrained.
. . . . . The comparative immunity of females of the higher
classes in hot climates from spinal distortions, may, in part,
depend upon their freedom from the pressure of stays and
bandages. Mr. Shaw has some good observations on this subject : 'It is, perhaps, correct to say, that the less exercise a
child takes, the more does she require general muscular relaxation in the recumbent position ; and, that the lighter and more
sedentary the pursuits are, the more necessity there will be
either for active exercise or general relaxation. Thus, in warm
climates, where active exercise cannot be taken, the due relation of parts, or balance of the system, is preserved by great
indulgence in the recumbent position.' "

On the theory of exclusive muscular action, in maintaining
the erect posture, a soldierly attitude should be the one which
all persons would choose while standing, for the more nearly
perpendicular the spine is kept, the more easily must the weight
of the head and trunk be balanced. Mr. W. Adams, has a
theory of "vigilant repose," the lateral muscles of the spine
while standing erect, being supposed to be in a state of relaxation, but on the watch, and just ready to act upon the least
deviation of the body from the plumb line.* Yet this is contrary to all experience, for soldiers themselves, as soon as they
are released from discipline, immediately assume attitudes in
which the spine makes first one set of ligaments tense, and
then the other. It is not so much to rest the muscles of the
legs, that the weight of the body is thrown first upon one and
then upon the other, as to rest the muscles of the spine, by
throwing the burden of balancing upon the ligaments.

From all this, the indication is plain, to give the muscles

---

* Lectures on the Pathology and Treatment of Lateral Curvature, and other
forms of Curvature of the Spine, by Wm. Adams, F.R.C.S., Surg. to the Royal
Orthopedic Hospital. Churchill & Sons. London. 1865.

variety of exercise, with frequent and abundant rest in the horizontal posture, saving the ligaments from any persistent strain, and, in confirmed cases, to strain the ligaments in the opposite direction, till they acquire equality of development on the two sides.

It used to be advised, in this class of cases, to carry a bag of sand, or other considerable weight, upon the head, to induce the muscles to hold the spine erect, as the easiest way of sustaining the burden. If the muscles fail to appreciate this advantage, in sustaining the weight of the head and shoulders alone, it is difficult to perceive how they should, when the weight of a bag of sand is added. The truth probably is, that in both cases, the muscles, as soon as they are fatigued, attempt to shirk the burden, by allowing the spine to settle to the extent of its lateral flexibility, throwing the strain upon the ligaments. Thus the greater the weight, the greater must be the strain, and the greater the consequent yielding and gradual aggravation of the curvature.

The system of movements devised by Dr. Ling, of Sweden, and, with more or less modification taught by Dr. Lewis and others, is well suited to prevent this class of deformities, and to correct them in their early stages. It is a happy reform, to introduce it into schools as a part of the regular daily exercises, not only for the purpose here considered, but to secure better general muscular growth, better digestion, and more balanced developments.

In unconfirmed curvature, from muscular weakness, an important means of giving vigor to the muscles is friction upon the skin over them. The efficacy of this, in imparting muscular tone, is well enough understood by every groom, and it is a pity that it should be so much neglected in human hygiene. Once, or oftener, in the twenty-four hours, the patient should lie upon her face, with the whole length of the spine exposed, when passes should be made slowly, and with a considerable degree of pressure, from the occiput to the sacrum, the passes all being toward the sacrum, and continued fifteen minutes at each period. It is convenient to practice this friction just

before bedtime, for its additional effect in procuring good sleep.
To give greater adhesiveness to the surface of the hand, it may
be moistened with some alcoholic preparation.

4. Lateral curvature, from spasmodic action of muscles,
finds its best illustration in torticollis, or wryneck, in cases in
which rigidity of the sterno-cleido-mastoid muscle is associated with spasmodic action of the spinal muscles of the same
side. Fortunately, the cases of curvature of this class are
uncommon in the middle and lower portions of the spine,
though the twisting of the neck and tilting of the head, from
the action of the sterno-cleido-mastoid, are common enough.

Figure 31, from Bonnet, illustrates a marked degree of
wryneck. These cases usually arise, like strabismus, from
sympathetic disturbances of nervous function, and the most
successful treatment is the early removal of the disturbing

Fig. 31.

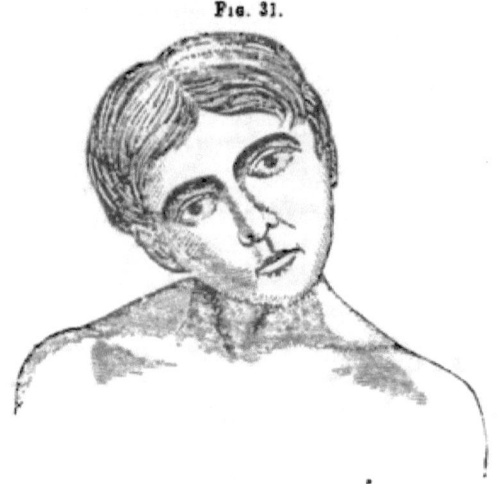

Torticollis, from Bonnet.

causes. The permanency of the deformity depends upon the
hypertrophied and shortened condition of the muscles of one
side, after the irritating disturbances have passed away, or upon
the condition of *contracture*.

Where these irritations have an hysterical character, the

effects may be expected to be more transient, while contractions resulting from the reflection of influences from sources of irritation, having a degree of permanence, must be less promising; and those resulting from irritation at the origins of the nerves in the brain or spinal cord, must be most unpromising of all.

The treatment in these cases must, obviously, have reference to the removal of the irritating cause, whether acting as an incitant of hysterical perversities, by reflex influences, transferring the result of irritation from the afferent nerves in another situation, to the efferent nerves going to the muscles affected, or existing at the origins, or in the course of the nerves supplying the muscles which are producing the disproportionate contraction.

The primary treatment must evidently be medical, but the permanent results may require systematized movements and mechanical appliances, partly to assist the elongated muscles of the convex side in regaining their volume and shortening their length, and partly to tire out and lengthen the contracted muscles.

The division of the rigid muscles may be necessary. By this means, time may be gained for righting up the spine, by means of lateral pressure, securing the commencement of change in the nutrition of ligaments, shortening those on the convex side and lengthening those on the concave side, so that when the cicatrization of the divided muscles restores their functions, the antagonist muscles may have been developed for effectual opposition.

The result of this operation upon the spinal muscles is better explained, in many instances, by the supposition that a profound impression is produced upon the nerves, analogous to that of the moxa, the actual cautery, and the galvanic puncture. This theory, however, would always lead to the employment of the latter remedies, rather than the division of the muscles. In the division of the sterno-cleido-mastoid muscle, for wryneck, however, more may be accomplished; for, by a free suppuration, a connecting cicatrix may be obviated, or if formed, it may be too thin to draw the two divided ends of the muscles together, especially if the antagonizing muscles are aided by appropriate apparatus.

All the other muscles concerned in producing lateral deviation of the spine must completely reunite after division, whether the incision is open or subcutaneous; and with all these muscles the division must prove a failure, if the disproportionate contraction of the muscular fibres is renewed after cicatrization. An element of the rationale of success in some cases of myotomy, practised as a remedy for muscular contraction, is the break in the succession of the irritation arising in the spasmodically contracted muscles, and reflected upon themselves.

In confirmed wryneck, in which the malposition has been permitted to go uncorrected during several years of the growing period, the sterno-cleido-mastoid muscle affords such resistance to restoration, that time and annoyance are saved both to patient and operator, by its division.

Unlike the case of the division of the tendons moving the foot, the function of the muscle can be compensated by the action of other muscles; and the absence or incompleteness of its reunion is a desirable result.

The following figure (Fig. 32) illustrates a mode of preparing, without the aid of an instrument maker, an apparatus for retaining the head in any desired position, connected or not with any operation.

The letters apply to the same parts in the two figures. The whole frame is made of hoop and sheet iron.

This simple, extemporaneous arrangement, modified to suit the case, will be found equally useful in rendering the head immovable after plastic operations upon the neck, where immobility is essential to the adhesion of the cut surfaces.   ·

It is not enough to counteract the tendency of this muscle to shorten; for, as in the case of talipes, the protracted distortion is accompanied by the shortening of many muscles which . cannot be divided, and by the elongation of others. A lateral curvature of the cervical portion of the spine is the necessary consequence, and the result cannot be satisfactory, until it is perfectly comfortable to the patient to be forced by the apparatus to carry the head in a position the opposite to that which was assumed before the beginning of treatment. It is only this thorough breaking up of the habit of malposition which will be satisfactory in the result.

Fig. 32.

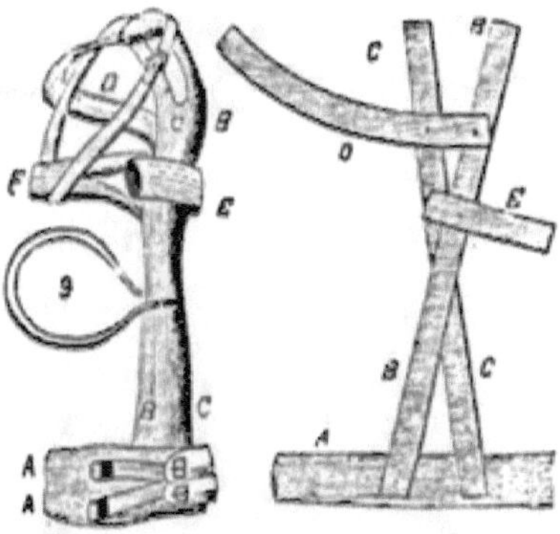

A. Base to encircle the pelvis.
B B, C C. Uprights made long enough to go over the top of the head.
D. Head-piece, made long enough to pass a little more than half around the head horizontally.
E. Neck-piece, to pass under the mastoid process and base of the jaw of the opposite or prominent side.
F. Chin piece, made of pasteboard, and covered with muslin, with its fastenings.
G. Shoulder loop, which sustains the weight of the apparatus upon the prominent shoulder.

It is quite possible, that a great degree of perseverance and attention may be adequate in almost all cases to overcome the deformity, but the progress is so much more rapid with the division of the muscle, and the operation, when properly performed, is so devoid of danger, that most surgeons will continue to follow the easier and more speedy method.

Mode of Division of the Sterno-cleido-mastoid Muscle.—The chief requisite is familiarity with the anatomy of the parts.

An incision is made on the outside of the muscle just above the clavicle, sufficiently large to admit a long probe-pointed bistoury, supported by a strong handle. This is carefully pushed along behind the muscle, until it has traversed the whole breadth of the muscle, its flat surface having been kept toward the muscle, when the bistoury is turned with its cutting edge toward the muscle, and a careful division is made by the pressure of the finger of the other hand pressing the muscle upon the blade of the bistoury through the medium of the skin.

As the muscle is made tense by the position given to the head, by the hands of an assistant, it is very easy to determine when all the fibres have been divided.

As the loose character of the areolar tissue will almost certainly prevent adhesion, it is not important to seal the opening with great care.

Figure 33, from Bonnet, exhibits a yoke resting upon the shoulders, for the purpose of supporting a brace upon each

Fig. 33.

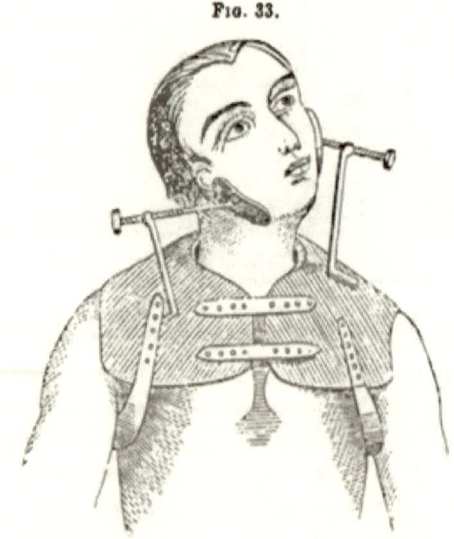

side, with a pad pressing upon the side of each jaw, adjusted by a screw. There is supposed to be a rest behind to keep the head from sliding backward.

Figure 34, also from Bonnet, exhibits an arrangement for suspending the head and affording some degree of motion.

A jacket with a metallic framework, only the front fastenings of which are seen in the cut, sustains a curved rod passing up behind and terminating above the head, from the end of which the suspensory apparatus depends.

FIG. 34.

If the attachment is made by elastic rubber, instead of metal as shown in the cut, a very considerable movement is secured, while the extension is still sufficiently strong. Something like this can be very easily extemporized.

TREATMENT OF CONFIRMED LATERAL CURVATURE.—The words of Lionel J. Beale, in his work "On Deformities," written more than thirty years ago, are peculiarly appropriate: "The great difficulty of treating this class of spinal distortions (lateral curvature) consists in the necessity of removing the weight of the body from the vertebral column, and at the same time using such exercises as will remove the evil by strengthening the muscles. . . . . . . Every day's experience proves to me the utter impracticability of removing

a decided lateral curvature of the spine, without mechanical
extension. It would be as easy to reduce a dislocation of the
femur without the assistance of art, as to correct a long-continued deviation of the vertebral column without extension.
In the earliest stages of the malady, exercise alone will
remedy the evil, but we are seldom consulted until the curvature is confirmed and permanent; that is to say, it cannot be
removed by any exertion of the patient, nor does it, as in the
earlier stage, disappear while the body is in exercise and the
muscles in action. If a child with incipient lateral curvature
is watched during the time when she is engaged in any active
and absorbing amusement, there will be no appearance of distortion, and at this period judicious management will prevent
the establishment of a permanent curvature. But when the
deformity has existed for a year or more, when it has become
fixed in any degree, the ribs will have accommodated themselves to the deviation, and they will contribute to the main-
tenance of it; in these cases no variety of exercise, no kind of
gymnastics, will do more than prevent further mischief. To
remove that which has become established, extension alone will
be of advantage. The judicious combination of extension,
exercise, and repose, will in time remove lateral distortions of
the worst kinds, but without unremitting perseverance they
will be of no avail, for it is not in a few weeks that good can be
effected in such cases, months and even years are required.
. . . . . . The time required for the cure of spinal deformities,
of course varies. During the period of growth, perseverance
will almost invariably insure success, and if the distortion is
recent the shape will be restored in a few months; but the
exercises must be continued *afterwards*, to prevent a relapse.
Even when the growth has ceased, if no absorption has taken
place in the bones, if they have not yet been rendered cuneiform by pressure, we may sometimes succeed in removing the
deformity entirely; but it is obvious, that when the osseous
system is much implicated, a complete cure cannot be effected,
and some degree of deformity must remain for life."

Dr. Beale's theory has been carried out by improvements in

appliances by Dr. H. G. Davis, whose language in a paper published in the Boston Medical and Surgical Journal, vol. 46, No. 5, p. 96, for March, 1852, is here quoted:

"After several years' experience in treating curvatures by apparatus and by various modes of exercise, the conviction was forced upon me, that in order to do it with any degree of assurance, it was necessary that the apparatus should be so contrived, that it would not only remove the deformity, but it should at the same time leave all the muscles free to act, and the patient under the necessity of using them to balance and support the body as fully as without the aid of such appliances; also, that it should be so planned that it should effect as much during the sleeping as the waking hours. . . . . . . For the furtherance of the recovery, the instrument should not be fixed or stationary in its adjustments, but should possess elasticity, so that if the form yields, the apparatus may follow up the change, and exert nearly as much force upon the curve as it did before it had yielded in any degree."

The introduction of elastic rubber was early seized upon by Dr. Davis, in great part to meet this necessity for elasticity, to which he gave the name of "artificial muscle."

The *desirableness* of elasticity in the appliances was appreciated long ago. The late improvements have been made rather in the achievement of the desideratum than in the conception of the want. Yet, up to this time, nothing has been devised which will do away with the necessity for the horizontal position in the treatment of confirmed cases. It is indeed probable, that when time enough can be devoted to a patient, the hands of the operator and his assistants, with some very simple appliances, will do as well as expensive and complicated apparatus. In this direction are the "antiplastic" movements of Werner.

"The patient is placed upon a covered table in a dorsal position; the hands of the operator are employed to correct the deformity, and to bend the spine over in the opposite direction; and, in fine, the patient is directed to maintain the same for an hour or so at a time. Another competent person may take the place of the physician to facilitate proceedings, which should be repeated several times during each day."

Mechanical Appliances.—There are two classes: those which produce direct extension, and those which act laterally upon the spinal column. The simplest of the first class is Darwin's bed, in which the head of the bed is elevated about twelve inches, and the cap investing the head is fastened to the headboard. The tendency of the body to slide downward makes the extension.

The annexed cut (Fig. 35), taken from Bigg's "Orthopraxy," illustrates a bed with its appliances for elastic extension, combined with elastic lateral pressure. The coiled wire shown in the cut, can more conveniently be replaced by elastic rubber.

Fig. 35.

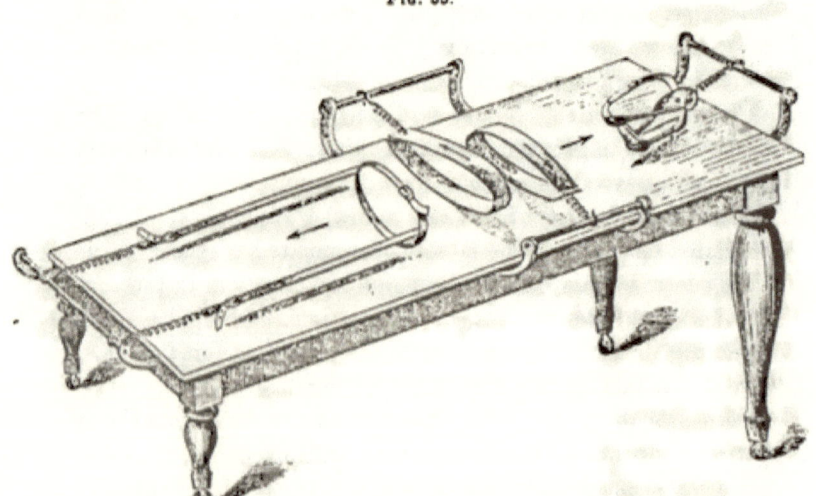

The plan in the figure contemplates extension by force applied, not trusting to the weight of the body on an inclined plane.

Brodhurst,* speaking of this apparatus, says:

* Curvatures of the Spine, &c., by Richard Brodhurst, F.R.C.S., &c. Churchill. London. 1864.

" The most effectual mode of treating spinal curvature is on the spinal couch. In the recumbent posture, the weight of the body is entirely removed from the spine, and the means are, therefore much more efficient to remove distortion than when portable instruments alone are employed. Not only is the recumbent position most favorable for the action of the instrument, but the instrument itself when attached to the couch, is infinitely more powerful."

It is hardly necessary to say that the force applied should not at any time be greater than can be borne without suffering.

The accompanying cut (Fig. 36), quoted from Delpech, by Beal, in his work "On Deformities," has an historic interest as showing a power and complication of machinery conceived in disregard of the limitations of human endurance. Concealed within the frame of the bed are pulleys, wheels and springs innumerable, to effect the simple purpose of extension.

FIG. 36.

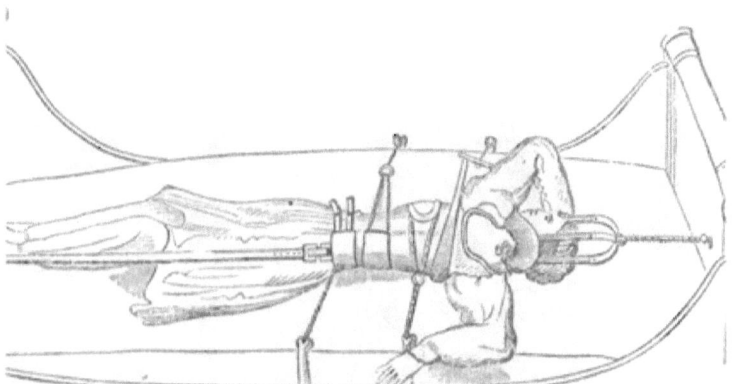

The appliances for lifting in the erect posture rest upon the pelvis, and receive the shoulders upon crutches; and where the head is to be supported a collar applies to the base of the skull and of the lower jaw.

This plan of mechanism was adopted by the older orthopedists, and lately, Dr. Edmund Andrews, of Chicago, has published plans of apparatus upon this principle, moulding a sheet of copper over an accurate cast of the pelvis, and extending a rod up behind the spine, and by means of attachments to this rod, lifting upon the head or shoulders, or both.

This plan is adequately illustrated by the following figure (Fig. 37), and explanation.

FIG. 37.

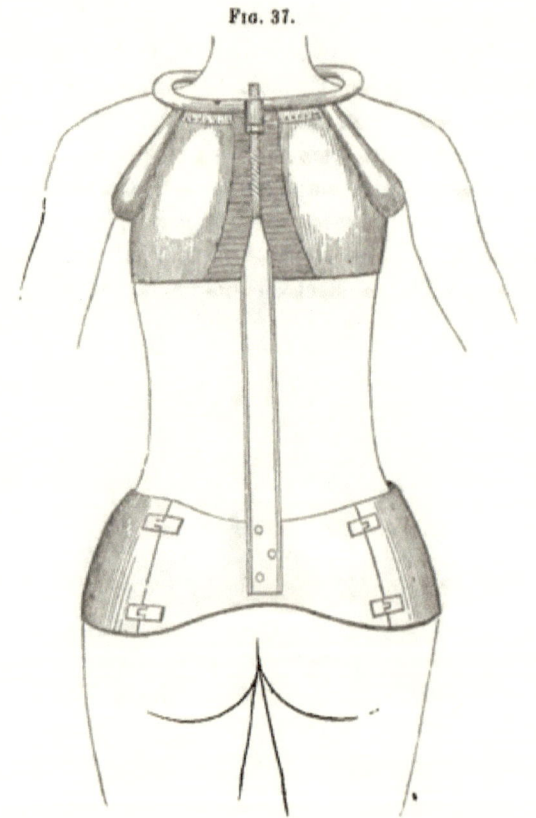

"If the disease is inflammatory and is not higher than the sixth dorsal vertebræ, I often make use of an apparatus upon the plan shown in Fig. 37, which is constructed in the following manner: First take a complete cast of the patient's hips in plaster of Paris, from the small of the waist downward to two inches below the trochanter major. Using the cast as a pattern, have a brass armor hammered to fit it, making it wide on each side, somewhat narrower behind, and still narrower in front, so that the thighs may not press against the lower edge when flexed. This armor opens by hinges situated a little external to the sacro-iliac junctions, and locks in front on the linea alba. It is, therefore, composed of three pieces; and, when clasped upon the patient, will be found to fit the hips nicely, and to bear any amount of downward pressure, without causing pain. It should be lined with cotton flannel. In children, in whom the hips are usually narrow, it is commonly necessary to modify this by making the side pieces in such a form as to cap over more decidedly the crests of the ilia. A steel rod arises from the centre of the back of the armor, and another from the front, each coming well up to the height of the shoulders. Their upper extremities are cut for eight inches into a screw, and carry an octagonal nut. A short and strong jacket of leather must be made to fit the chest snugly, and fastened at the top to the circumference of a steel ring which surrounds the neck. This ring has sockets before and behind, which slide down upon the screws to a distance regulated by the nuts. A cross arm before and behind may be used instead of the ring. If now the two nuts be screwed upward, the ring will be raised, and by the tension upon the jacket the weight of the upper half of the body may be entirely taken off from the spinal column, and borne by the steel rods directly upon the armor of the hips. The source of irritation being thus removed, the inflammation will, in many instances, subside spontaneously without any other treatment. At the same time, the spinal column is drawn straight, exactly as if it were a string. The jacket may be advantageously lined with adhesive plaster in some cases. If the seat of the disease is not above the middle of the dorsum, the very excellent corset of serpentine wire, devised by Dr. Wood, is often the best appliance."

Hossard is said to have been the first to attempt to support the spine by apparatus fitting the trunk, and producing lateral pressure and counter-pressure, while the patient is in the erect posture and walking about. This, **and the modifications of it, are known as** Tavernier's Belt.

Fig. 38, from Tamplin, illustrates a transition or combination.

FIG. 38.

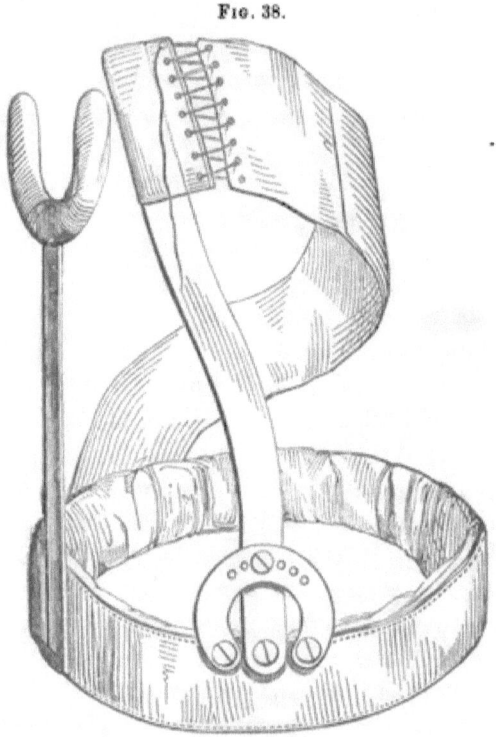

(From Tamplin.)

The combination is also preserved in the following figure (Fig. 39), modified from Bigg (Orthoproxy), in which the crutches are retained and lateral pressure is maintained by pads, each of which is placed upon an upright, the position of which is regulated by a ratchet at the base, controlled by a key. The lacing band in front and the shoulder-straps are not shown in the figure. Mr. Bigg thus describes the method of applying the apparatus:

"Stand behind the patient, and opening the pelvic band,

place it firmly around the hips, in such a manner that the arms rest upon the crutches; see that the two plates rest gently against the arc [convexity] of each curvature, the vertebral levers having been expanded previously to placing the instrument on the patient's body. Fasten the lacing bands in front, and then gradually tighten the vertebral levers by means of the key. Lastly, see that the arm-slides are of such a height as to maintain the shoulders parallel with the pelvis, and fasten the shoulder-straps. The instrument can rarely be worn more than four hours on the day it is first applied, but after three days the patient submits to it, and often feels greatly disinclined to part with the apparatus."

Fig. 39.   Fig. 40.

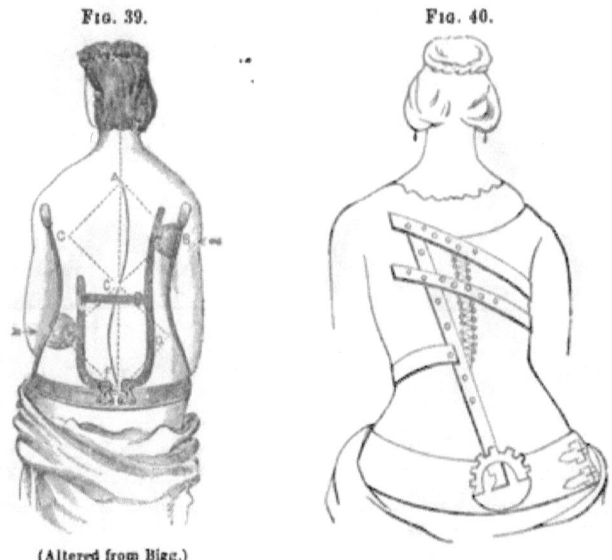

(Altered from Bigg.)

The alteration from Bigg consists in representing in the figure an elastic rubber connection between the two uprights, or vertebral levers, so that one of them being fixed and the other free at the base, a greater pressure may be endured from its yielding character.

This modification of Tavernier's belt (from Bauer; Fig. 40), applies the principle of lateral pressure alone. The reader

must suppose an upright in front to receive the other ends of
the straps.

The next figure (from Andrews, Fig. 41) exhibits another
form of applying the same principle. In this apparatus, the
pelvic piece is fitted to the pelvis with great accuracy, employing sheet copper for the purpose, in order to afford some stead-
iness to the upright.

FIG. 41.

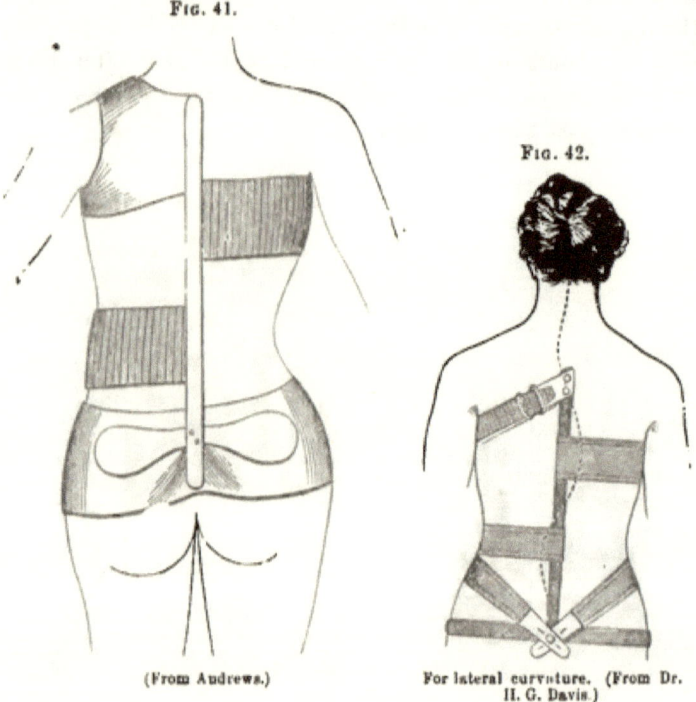

(From Andrews.)

FIG. 42.

For lateral curvature. (From Dr.
H. G. Davis.)

The accompanying cut (Fig. 42) illustrates the plan of Dr.
Henry G. Davis, of New York, which combines lifting with
lateral pressure, and admits, from the elasticity of its rubber
bands, of very considerable movement of the trunk, while the
elasticity of the material tends all the time to remove the
lateral deviation.

The lifting element is obtained by the oblique bands passing
over the crests of the ilia, which are also elastic, thus avoid-

ing the necessity for any very nice adjustment of the pelvic band.

The figure shows four locations of pressure, corresponding with the extremities and the two convexities of the sigmoid flexure.

The straps are supposed to be attached to an upright in front, similar to that behind, curved to correspond with the form of the trunk.

Chairs have been constructed for the same purpose, a very perfect, though complicated specimen of which is this (from Bonnet) Fig. 43.

FIG. 43.

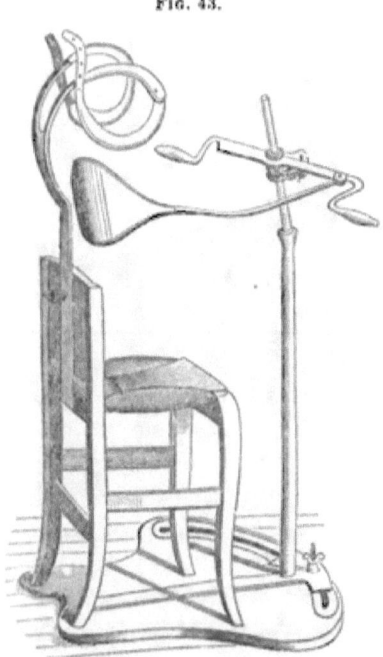

Chair for lateral curvature (from Bonnet).

An attachment coming up from behind holds the shoulders immovable, while the pad is applied to the ribs on the projecting side.

Fig. 44, p. 120, shows the patient in position. He has the

power of regulating the degree of pressure upon his ribs, by his hold upon the lever in front of him.

Fig. 44.

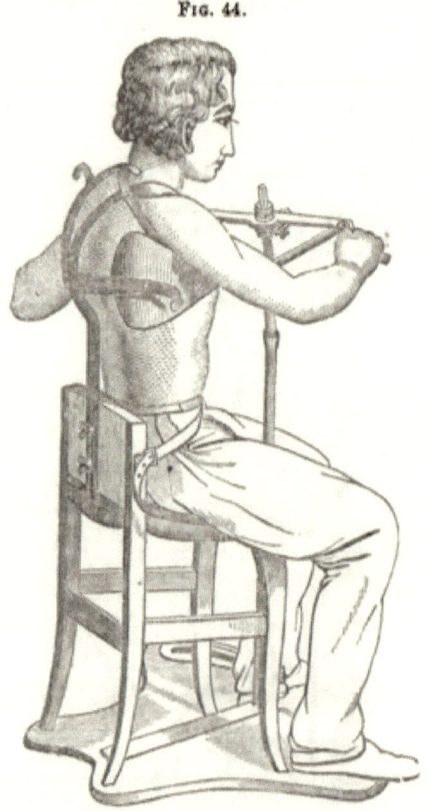

Chair with its sitter (from Bonnet).

A more simple, and apparently more useful apparatus has been devised by Dr. Andrews (Fig. 45, p. 121), in which there is counter pressure upon different portions of the spine, upon the two sides.

The patient is tempted to aid the influence of the pressure by muscular action, to escape the approach of the pads.

Fig. 46, p. 122, is adapted to more obstinate cases.

Four locations of pressure are secured, corresponding with

the extremities of the sigmoid flexure, and the two intermediate convexities.

Any one who visits a "movement-cure" establishment, like
that of Dr. C. F. Taylor, of New York, will see a variety of
ingenious contrivances to answer the purpose of these chairs.

FIG. 45.

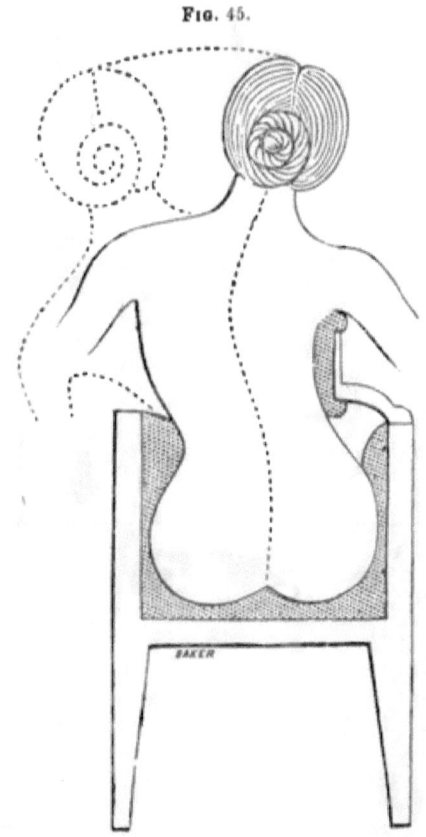

Andrews's chair for exercising the deficient muscles of the spine.

For the convenience of those who live beyond the reach of
skilled instrument makers, I have made an endeavor to devise
a plan, which can be executed by any ordinary mechanic, and
by any physician who will give his time to the work. The

framework of this apparatus is sufficiently illustrated by Figure 47, p. 123.

FIG. 46.

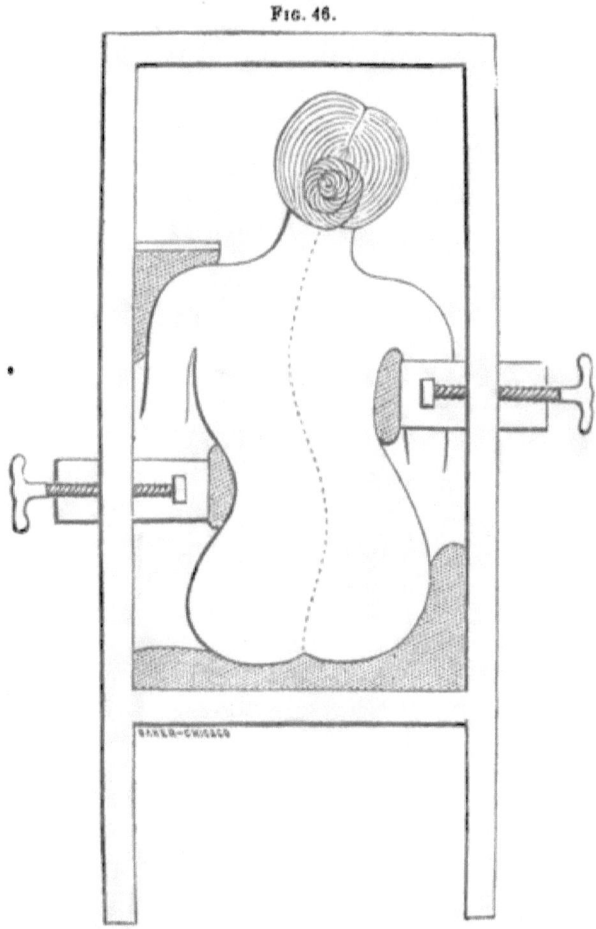

Explanation.—A metallic frame for lateral curvature, which may be extemporized.

a a. Two uprights of hoop iron, or, better, of thin steel. *b b.* Base of copper, tin, or sheet iron, reaching two-thirds around the pelvis, to which the uprights are attached. *c c.* Two pieces of hoop iron, or of thin steel, for the side on which the shoulder is elevated. *d.* A thinner and narrower piece of steel, to pass under the armpit of the depressed shoulder. This is movable at the rivet attaching it to the uprights, and upon the distal end is attached an elastic strap, which passes up in front and over upon the prominent shoulder, and down to a buckle attached to the base *b b*, in order to make the lift under the depressed shoulder, and the weight upon the opposite elevated shoulder balance each other. *e.* A strip of hoop iron (or other soft iron) to apply to the neck and base of the skull. All parts are properly padded and armed with straps and buckles.

To retain the head, after division of the sterno-cleido-mastoid muscle for wryneck, and after plastic operations upon the neck, a frame, upon the same principle, can be easily made, prolonging the uprights *a a* to pass over the top of the head, and employing two neck-pieces, *e.* (See Fig. 32, page 107.) The uprights for this purpose are made short at the lower end, so as to bring the base *b b* around, the lower part of the chest. The pieces *c c* and *d* are left off.

Fig. 47. Fig. 48.

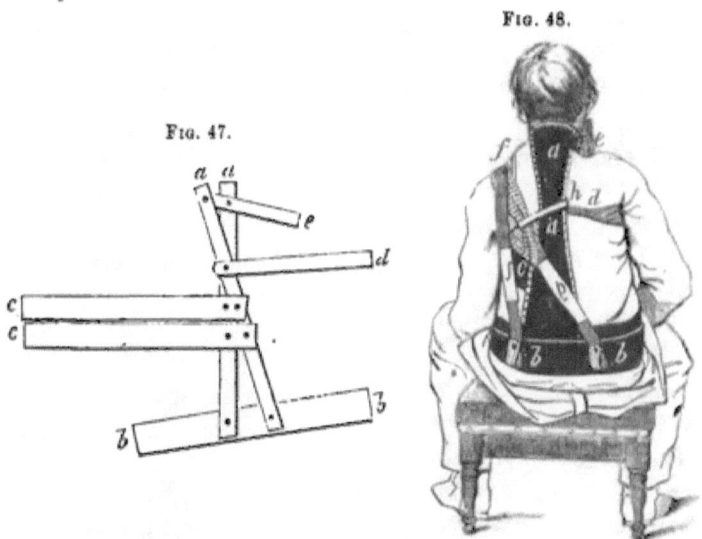

In Fig. 48, the apparatus is seen applied. It will be

observed that it differs from the preceding plans, in making the prominent shoulder the centre of force, and, in the erect posture, the support, also, of the weight of the apparatus.

Fig. 48. Apparatus for lateral curvature applied. *a a*, *b b*, *c*, *d*, *e.* The same as in Fig. 47, properly bent to the form of the body and padded. *ff.* An elastic strap attached to the front or distal end of the bar *d*, passing up in front, and over the opposite shoulder, overriding the loop *g*, and passing down behind to the base *b b*, where it is attached by a buckle. *g g.* A loop encircling the prominent shoulder, and attached by an elastic strap to a buckle fastened to the base *b b*. *h.* A small cord passing around one of the branches of the loop *g*, and attached to the upright, for the purpose of keeping the loop from slipping off the shoulder. By means of the strap *ff*, and the loop *g*, the apparatus is made self-sustaining, and it is not necessary that the base *b b* should have any nice adaptation to the pelvis.

The apparatus, as shown in the figure, is adapted to a curvature in which the left shoulder is prominent. The bar made into a hook *e e*, and the right extremity of the pelvic base *b b*, also made into a hook, passing around the right ilium, are intended to be nearly unyielding. The bar *d* moves with the ascent and descent of the right shoulder, and all the straps and fastenings are elastic. This is to give the muscles more freedom of motion, both for comfort and to permit that to-and-fro movement, which gradually secures yielding of both muscles and ligaments, without danger of inflammation or ulceration anywhere. A vastly greater amount of pressure can be borne with this yielding elasticity, than without it.

The details of the apparatus admit of a great variety of modifications, to suit different cases. If it is desirable to support the head, to give the cervical vertebræ some extension, the weight of the head may be supported upon the bar *e*, and one placed upon the other side to correspond with it. The counter-extension thus comes to apply to the projecting shoulder, through the strap *f*, and the loop *g*, which pass over it. It is believed that this plan admits of a greater variety of application than any other which has been recommended, at the same time that it can be made from the most common materials, requiring very little skill.

The employment of the best steel and the best workmanship, to make the whole apparatus light and elastic, is desirable, but not necessary to success.

Plans of apparatus for the treatment of lateral curvatures of the spine migh be multiplied, but those here shown are sufficient to illustrate the principles involved.

What has already been said of the need of patience and perseverance in undertaking to remove deformities attendant upon confirmed curvatures and twistings of the spine, with increase of convexity of the ribs on one side and diminution on the other, is sure to be appreciated by those who make the attempt.

Fig. 49 illustrates a case belonging to the second division of the classification here adopted —lateral curvature, occurring in a lad seventeen years of age, of previous good health. The deformity has been chiefly produced during the last six months, and is observed to be progressing rapidly.

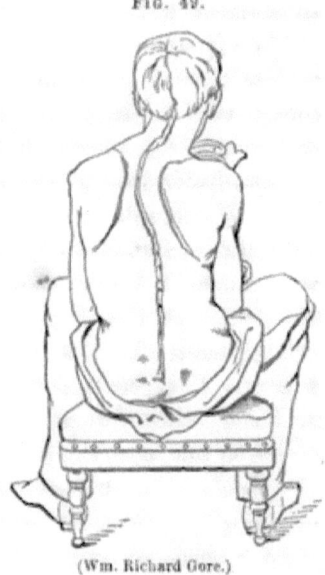

Fig. 49.

(Wm. Richard Gore.)

He fell from a horse, three years ago, striking upon his head, and was nearly helpless for a few days after this. This is the only known cause. There is very little deviation in the lower dorsal, and in the lumbar vertebræ. Stature five feet.

After ten days' use of apparatus delineated in Figs. 47 and 48, the patient's height had increased half an inch.

In addition to mechanical treatments, the patient takes,

three times a day, a teaspoonful of syrup of iodide of iron and
simple syrup.

Aug. 3d, 1864 (about two months from the commencement of
treatment), increase of stature since last measurement, $\frac{5}{8}$ inch
—whole increase, $1\frac{1}{4}$ inch. Nov. 5. Additional increase of
stature, $\frac{3}{8}$ inch—total increase, $1\frac{1}{2}$ inch. At this point, there
seemed to be a cessation of supposed softening of bone, and a
fixedness of position accrued, so that no further progress was
made. The general health improved. Perhaps greater amendment of form might have been secured, by requiring the patient to keep the horizontal posture, but it seemed necessary
to permit him to move about, in aid of his general health.

To the medicine is greatly due the credit of the arrest of
the softening of bone, but the mechanical support retrograded
the deformity before the medicine had time to act. At a later
period, the restoration of bony firmness rendered further
amendment of form impossible.

The variety of lateral curvatures to which this case belongs,
is closely allied to the antero-posterior curvatures, next to be
considered. Inflammation and softening, or destruction of the
substance and connections of bones, occur alike in both.

The involvement of the articulations of the oblique processes of one side of the spine, permits a lateral deviation,
while the escape of these articulations implies a curvature forward, as the bodies of the vertebræ and their intervertebral
fibro-cartilages yield.

*Antero-posterior, Vertical, or Angular Curvature—Pott's
Disease—Kyphosis.*—This curvature depends upon disease of
the bodies of the vertebræ, resulting in destruction of the substance of the bone, while the articulations maintain their in-
tegrity. As the weight of the superimposed parts forces the
adjoining surfaces of the vertebral bodies together, the spinous
processes must separate and project backward, with a more or
less sharp curvature.

The pathology does not differ from that of inflammations and
degenerations existing in other spongy bones, either as chronic
sequels of inflammations following injuries, or as degenerations

of spontaneous origin; the localizations of constitutional ten-
dencies, being directed to this seat by any cause diminishing
the proportionate vigor of nutrition of the spinal bones. The
freedom from acute pain, probably, depends upon the low grade
of inflammation giving time for a slight expansion of tissue,
to accommodate the increased amount of blood, and saving
those nervous extremities from distension, which, when irritated, are susceptible to painful impressions. A dull, ill-defined,
aching sensation, instead of an acute pain, usually attends the
early history of the cases.

A careful step to avoid jars of the body, and a slightly stoop-
ing posture, with a disposition to brace up the **trunk by placing
the hands upon the thighs, and a careful avoidance of all those**
attitudes which **bring increased** weight upon **the bodies of the
vertebræ, are characteristics which mark the progress of the**
disease. **In picking up objects on** the floor, the patient carries
his hand to the floor by flexing the hip- and knee-joints, holding
the spine stiff and nearly erect; a movement so contrary to
that of persons with healthy spines, as to attract attention.
Rheumatism may occasion constraints of movements very similar, and hence a particular caution is necessary in distinguishing ostitis of the vertebræ, until the persistence of the **symp-**
toms renders the diagnosis unequivocal.

From the movements of the patient, we may see that there
is an effort to save the vertebræ from **pressure, though the sensations** are so diffuse that the patient **is not aware of the seat**
of the uneasiness. Those startings and spasmodic muscular contractions which attend inflammation of **bones adjoining** more
movable joints, are usually absent, owing to the absence of the
friction of rough surfaces, such as **apply to each** other in
destructive inflammations of **the knee- and** hip-joints. The
patient is able to save the bodies of the vertebræ and the intervertebral cartilages from pressure while in the erect posture, by
active contraction of the erecting muscles of the spine attached
to the spinous processes, but they soon tire out and yield to the
tendency of the spine to curve, just as muscles elsewhere yield
to an overcoming force applied by extending apparatus. This

yielding becomes habitual and permanent with increased length of muscles. As this process goes on, the ligamenta subflava connecting the arches of the vertebræ become permanently elongated by change in nutrition, while the articulating processes assume new relations with each other. An abrupt projection thus occurs in marked contrast with the curves of the lateral deviations. No tenderness on pressure usually attends this process of destruction, as the diseased parts are beyond the limit of the influence of pressure upon the spinous processes, upon the ribs, or upon the adjacent muscles.

It will be interesting to quote in this connection, from Thomas Copeland, who wrote half a century ago ("Diseased Spine," London, 1815):

"Most of the prominent symptoms of this complaint are derived from interruption of the nervous function in the parts below the seat of pressure.

"The debility and inaptitude for motion, early fatigue, and relief from rest, may partly result from the affected bone or intervertebral substance. But in most cases, and most particularly when the disease is in its usual place, the superior dorsal vertebræ, the great characteristic symptom and circumstance *is a commencing paralysis of the abdominal muscles*. It is surprising how very early in the disease this symptom may be detected, when the attention is directed to it. It is sometimes described as an oppression in breathing, tightness of the stomach, a band tied around the belly, torpor of the abdomen, and by other expressions in different patients.

"It produces costiveness and retention of urine in a more advanced stage. . . . . . . So early does paralysis of the abdominal organs present itself, as a symptom of compression of the spinal marrow, that it is very generally treated for a time for the complaint which it most resembles. I have seen it called asthma, dyspepsia, and even diseased liver, from the sense of uneasiness and stricture over the region of the liver and stomach; sometimes it is taken for a disease of the colon or rectum, from the accompanying costiveness and pain; the bladder also, being unable to perform its office, the cause of the impediment is sought for in the urethra and kidneys."

Mr. Copeland claims the discovery of the increased sensibility to the touch, and **to the** presence of heat in parts in the neighborhood of carious vertebræ. He speaks thus of temperature. "A sponge wrung out of hot water, and carried down along the spine, will often give a very acute pain, while passing over the part where the disease is going on. This I first discovered by accident. When I had been applying leeches to a diseased spine, the **gentleman who** was my patient, complained of great pain **when the sponge passed over** the projecting **vertebræ."**

From the intrinsic interest **of the subject, and the interesting manner in which** Pott, **from whom** the disease is named, **detailed the symptoms,** the attention of the reader will not tire in reading this somewhat lengthy quotation.

"When the disease attacks one who is old enough to have walked properly, the awkward and imperfect manner of using the legs, is the symptom which first attracts attention; and the incapacity of using them at all, which soon follows, fixes that attention and alarms the friends. The account **most** usually given is, that for some time previous to the incapacity, the child has been observed to be languid, listless, and very soon tired; that he **was** unwilling to move much **or** briskly, that he had been observed **frequently to** trip or stumble, although no impediment **lay in his way, that when he moved hastily, or unguardedly, his legs would cross each other in**voluntarily, **by which** he **was** often suddenly **thrown down;** that if he attempted to stand still **and** upright, unsupported by another, his knees would totter **and bend** under him; that he could not with any degree **of precision direct** either of his feet to any particular point, **but that in** attempting to do so, they would be suddenly and involuntarily brought across each **other;** that soon after this, he complained of frequent pains and twitchings in his thighs, particularly when in bed, and of **an uneasy** sensation in the pit of his stomach; that when he **sat on a** chair or stool, his legs were almost always across

9

each other, and drawn up under the seat, and that in a little time after these particulars, he totally lost the power of walking.

"These are the general circumstances which are found, at least in some degree, and pretty uniformly, in most infants and children; but there are others, which are different in different subjects.

"If the incurvation be of the neck, and to a considerable degree by affecting the vertebræ, the child finds it inconvenient and painful to support its own head, and is always desirous of laying it upon a table or pillow, or anything to take off the weight. If the affection be of the dorsal vertebræ, the general marks of a distempered habit, such as loss of appetite, hard, dry cough, laborious respiration, quick pulse, and disposition to hectic, appear pretty early, and in such manner as to demand attention; and as in this state of the case, there is always, from the connection between the ribs, sternum, and spine, a great degree of crookedness of trunk, these complaints are by everybody, set to the account of the deformity merely. In an adult, the attack and the progress of the disease are much the same, but there are some few circumstances which may be learned from a patient of such age, which either do not make an impression upon a child, or do not happen to it.

"An adult, in a case where no violence has been received, will tell you that his first intimation was a sense of weakness in his back bone, accompanied by what he will call a heavy dull kind of pain, attended with such a lassitude as rendered a small degree of exercise fatiguing; that this was soon followed by a sense of coldness in the thighs, not accountable for from the weather, and a palpable diminution of their sensibility. That in a little time more, his limbs were frequently convulsed with involuntary twitchings, particularly troublesome in the night, that soon after this, he not only became incapable of walking, but his power either of retaining or discharging his urine and fœces, were considerably impaired, and his penis became incapable of erection.

"The adult also finds all the offices of the digestive and

respiratory organs much affected, and complains constantly of pain and tightness at his stomach."

Notwithstanding the admirable clearness of this statement of symptoms, Mr. Pott, referring to the diagnosis of caries of the spine, says, " When these complaints are not attended with an alteration of the figure of the back bone, neither the real seat nor the true nature of the distemper, are pointed out by the general symptoms, and consequently they are frequently unknown, at least while the patient lives."

An end of no small interest in these quotations is the preservation of the precise form, as well as the substance of the knowledge possessed by those medical authors, upon whom we are sometimes tempted to look with contempt, because we have now some light which they had not.

The amount of destruction of bone is truly astonishing ; the bodies of several vertebræ in some instances disappearing. A striking case of this kind is given in Cruveilhier's Pathological Plates, fourth livraison, plate 4, in which there was a loss of the bodies of five vertebræ, the 5th and 11th dorsal vertebræ coming in contact by their anterior surfaces, the spinal canal making an acute angle, without the occurrence of paralysis at any time. The illustration of this case is copied in figures 50 and 51.

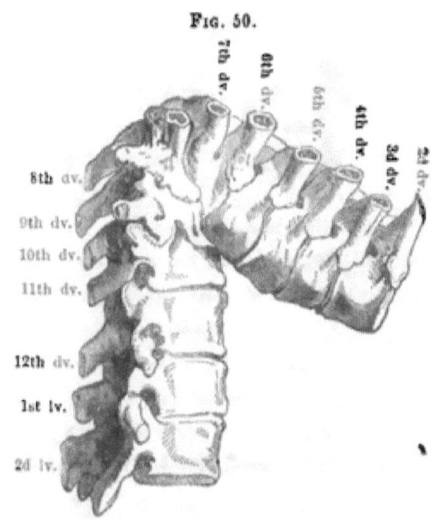

Fig. 50.

" The individual to whom this specimen belonged was not paraplegic, although the angle was so acute, but the pressure of the spinal marrow was probably prevented by the meeting of the eleventh with the fifth dorsal vertebræ, which furnished a groove for the reception of the eleventh. The bodies of the 6th, 7th, 8th, 9th, and 10th dorsal vertebræ have almost entirely disappeared, their confused remains forming a mass covered with osseous vegetations. The foramina were all preserved, but deformed and diminished in size. The spinous processes corresponding to the lost vertebræ had undergone a remarkable deviation; instead of projecting more than usual, they were much inclined and even slightly curved, so as entirely to fill up the interval which separated the 8th from the 9th, and the 9th from the 10th dorsal vertebræ, and to complete posteriorly the medullary canal. The manner in which nature had preserved the spinal canal in the midst of such devastation is remarkable. The vertebral column having been sawed lengthwise, we see (Fig. 51) with what apparent art the canal had been protected.''

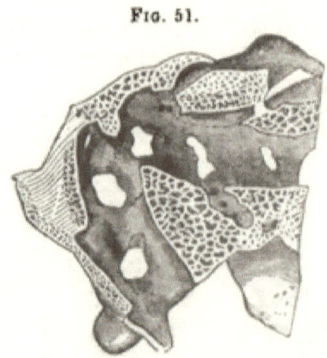

Fig. 51.

The occurrence of pressure upon the spinal cord is obviated by the slowness of the changes and the ultimate incasing of the cord by new bony material coincident with the anchylosis of adjoining vertebræ, except in those cases in which the weakened bony substances are fractured by the weight of the parts sustained, or by some movement or force suddenly applied greater than the vertebræ can endure. By this sudden fracture, the spinal cord may become compressed by the fragments and paralysis ensue. By the absorption or disintegration of these, or by their change of position through extension acting through the spinal column, the pressure may disappear, or through failure of these processes, the pressure and attending palsy may be permanent.

By looking at these illustrations taken from Cruveilhier, it will become obvious, that the explanation of the deformity is twofold.

1. As the substance of the bodies of the vertebræ softens and disintegrates during the inflammatory process, the weight of the parts above, while the patient is in the erect posture, and the contraction of the muscles on the anterior part of the body, overbalancing those upon the posterior (by the greater distance of the former from the spinal centre of motion), tend to compress the yielding bodies of the inflamed vertebræ, and to approximate parts **between which there is a hiatus** from loss of substance.

2. In analogy **with what is observed in other parts, a very** considerable **degree of deformity** may be supposed **to arise in** the **progress of** *restoration* after the destruction of a very **con-** siderable **amount of** osseous material, independently of any in-fluence **of weight or posture.**

The new-formed bone observed to have been produced in many specimens preserved in cabinets, must have had the properties of provisional callus, with a disposition to contract and disappear. In this process, there must be a tendency to bring the original parts between which this new-formed mate-rial exists, nearer together. **As the oblique or** articulating processes admit of no shortening **of the portion of the** spinal column, constituting the posterior **wall** of the spinal canal, the approximation **of** the bodies of the vertebræ, by this kind of cicatrix-contraction, must bend the spinal column, and **pro-ject the spinous** processes **backward. If,** for illustration, **the** amount of destruction of bone exhibited in the case quoted from Cruveilhier, can be supposed to have occurred independently of posture or pressure, there could have been no restoration of the strength of the spinal column, except by a filling up of the space with new-formed bone, which must have resulted in con-traction, up to the approximation of the original bone, above and below.

The duration of the disease, from its incipiency to the ter-mination by anchylosis, is measured by years. During the period of aggravation, pus may form and accumulate in **such** quantity as to burrow through the adherent pleuræ into the lungs, or along the fasciæ of the muscles into the lumbar region, or along the psoas muscle into the groin. Cases, however, may doubtless be cut short previous to the suppurating stage, while little or no deformity has ensued, as in similar inflammations in other cancellous bones and their investments. It may not then be apparent to the **patient or** friends, or even to the physician, what is the magnitude of the evil cut short.

There is a doubt thrown upon the efficacy of treatment in **this and other diseases** pursuing such a protracted course, be**cause the disease does not at once** begin obviously to ameliorate **upon** the institution of treatment. From analogy and **experience,** however, the following indications may be safely affirmed:

1. The disease, except when it is tubercular or a cancerous degeneration, has **an** acute stage, in which the treatment must **be** like that in the **acute** stage of other diseases of general low grade. This element of treatment is important before the bony substance is so absorbed or disintegrated that the pressure of **the** weight of **the parts above in the** erect posture can have any mischievous influence. Whether the morbid process, have their origin in an injury, **or in** some obscure constitutional influence localizing itself in the spine by an accidental selection, the case should be treated on the same principles as strumous inflammation in the eye. General heroic treatment cannot be borne, but active purgation once a week, cleaning out the colon and stimulating the liver and other alimentary glands, creating an appetite and giving a healthier hue to the surface, will constitute an important element in the treatment. Mercury is not an indispensable element in the purge, but it certainly adds very much to its efficacy. A grain of calomel for a child five or six years old, taken at night, and followed in the morning with an efficient senna draught, may be mentioned as an appro**priate purge.** The mercurial **may, for** convenience, be combined with a sixteenth of **a grain of tartar** emetic, two grains of leptandrin, and two grains of sugar, for a patient of this **age.** Where the mercury is omitted, two grains of leptandrin

may be given at night, followed by castor oil in the morning. Should there be considerable febrile excitement, the force of the circulation may be diminished by small and frequently repeated doses of tartar emetic, or in more active cases, veratrum viride. This element of treatment may be expected **to** be of most efficacy while the diagnosis is obscure, the disease being suspected by the gait and posture of the patient, before any deformity is manifest. It may be said in favor of the treatment that if the diagnosis is wrong, the remedies are equally appropriate for the rheumatic stiffness which simulates the spinal disease.

2. Following this **class of remedies, or in connection with** them, the **employment of iron and** iodine, with or **without other tonics, may be expected to have an important influence** in overcoming the constitutional perversion upon which the perpetuation of the **local** disease so much depends. The syrup of iodide **of iron, of** the United States Pharmacopœia, in doses of fifteen drops three times a day, for a child six years old, may be instanced as a convenient and efficient form of prescription. It is important to insist, that when there is any fulness of bloodvessels felt in the pulse, or seen in the capillaries of the surface, this class of remedies should be preceded or accompanied by a proper eliminative treatment. It is feared that the present prevalence of the humoral pathology in medical theories, too often leads to a neglect of this important principle of treatment to prepare the system for the absorption or tolerance of iron tonics.

3. **Local** remedies; cupping and leeching, emollient applications, the most convenient of which is a damp towel persistently worn over the spine, and moistened as often as it dries, slightly to reduce the temperature of the skin, and to soothe the nervous extremities, may be supposed to act chiefly by reflex influence, being applied to nerves having origins near to the origins of those going to the diseased parts, or identical with them.

Dr. Esmarch, applies dry cold to the spine in Pott's disease, and claims that great benefit is derived.

He first makes a mould of the back by means of a **sheet of** gutta percha, and from this, a metallic box is made with **one** side to fit the trunk. **There is an** opening through which **to** introduce cold water, and **there** may be an additional opening for the exit of water which has become warm, so that if need be a stream of water may be kept running through.

Dr. Esmarch thinks that moisture is often a source of hurtful **irritation** which is entirely **avoided by confining ice** or cold water, in water-proof inclosures.

Soft rubber **is** the material which he generally **prefers, and** some absorbing fabric should always be placed **between this** and the cutaneous surface, to absorb the moisture which **condenses** from the atmosphere.*

Blisters, moxas, and cauteries, however, if employed in this stage, should be applied at a distance, so as not to excite an increased sympathetic activity in the diseased parts. It is not impossible that **the disrepute** into which these last remedies have fallen is owing **to the disregard of** this therapeutic principle by the old surgeons, who applied these remedies too indiscriminately. Observing their efficacy **in the** later stages **of** the disease, **in** which spasmodic and painful muscular action had been sympathetically produced by irritation of **the nerves** of the diseased tissues, a hasty generalization may have **led** to their too early employment and to their application too near the diseased tissues. It is hardly possible that their employment should have been an entire mistake. **The extent** to which Brown-Sequard has insisted upon **this** latter point will doubtless have a controlling influence upon future **practice.**

A practical question will often arise, whether a purulent accumulation in the groin, or elsewhere, should be opened? **The** elements of the answer to this question may be thus stated:

Pus inclosed in the tissues **and** kept from exposure to the air, whether in large quantities or small, is capable of maintaining its original composition for a long period, and of being

---

* "The Use of Cold in Surgery. By Fr. Esmarch, M.D. Translated by E. Montgomery, M.D. Published by the New Sydenham Society, London, 1861."

finally absorbed.   Undecomposed pus is unirritating to the
tissues, does not interfere with healing processes in contact
with it, and travels chiefly by the pressure of its own weight,
or of its bulk when in too confined a space, as in phlegmonous
diseases, in which pus rapidly accumulates.   The presence of
undecomposed pus, therefore, will neither aggravate the disease
nor prevent the healing of the abraded surfaces of bone and
cartilage.   On the other hand, the complete evacuation of the
pus is impracticable, in consequence of the tortuousness of the
passages, or the formation of pockets from which the pus can
only be slowly drained by position.   The admission of air to
a portion of the pus within the tissues, sets to work a process
of decomposition which gradually communicates to the whole.
This inflames the interior of the pus-holding sac, and the
tissues originally diseased may take on an aggravated form of
inflammation from this new source of irritation.   The attempt
to evacuate the pus through a tube ending under fluid, can never
evacuate the less liquid portions of the pus, which are liable
to block up the tube, and if successful in avoiding the introduction of air into the cavity, the wound must heal by the first
intention, to avoid the unwelcome result of decomposition of
pus and consequent irritation.   If, however, an opening has
spontaneously formed into the air-passages of the lungs, or in
any inconvenient location, the case can be no worse by the
making of a free opening in the most dependent position, the
better to drain the parts, and when there is an opening into
the lungs, to save the air-passage from the burden of this discharge.   The introduction of local medicines by injecting
them into the purulent cavity, other than disinfectants, prom-
ises too little to encourage the practice.   ✦

  After the formation of pus has ceased in the progress of restoration, and after the parts originally diseased are well covered
by granulations, reducing the case to the condition of an ordinary chronic abscess, a state of things indicated by the greatly
improved state of the patient and the increased strength of
his back, there can be no more objection to the evacuation of
the pus than in an ordinary chronic abscess.

4. Quiet of the general system by confinement, chiefly in the horizontal posture, is of greater importance than parents can easily appreciate, both for obviating the general arterial excitement upon which the activity of the local disease in part depends, and for relieving the diseased parts from any functional action in sustaining the weight of the body. Whatever objections may be urged against the observance of the horizontal posture, in the later stages of the disease, in the period of wasting from suppuration, and in that of repair and progressing anchylosis, none can be urged on therapeutic grounds against the observance of rest and posture during the early period of subacute excitement, while tonics cannot be borne unless preceded by eliminants. Indeed, it is probable that if **this** indication could be properly appreciated by parents and physicians, in the forming period of the disease, vast numbers of cases would **recover,** without proceeding so far as to afford a clear diagnosis, and of course without any deformity.

The importance of rest **in the treatment of** caries of the spine, has a long time been appreciated. Copeland, writing half a century ago, quotes the clear language of Ford, which is better than any possible paraphrase **or** condensation. " It has happened to me so frequently to observe that the mode of treatment by absolute rest, confinement to bed, has been particularly successful in those cases in which a paralytic state of the lower limbs was added to the other inconveniences of the complaint, that I have been induced to think that this paralytic symptom, alarming as it is, conduces to the recovery of the patient. Without doubt, the paralysis indispensably compels the patient to a state of quietude in a horizontal position, **whereby the** pressure of the head is taken entirely from the distempered bones, and the establishment of a union between their ulcerated surfaces becomes thus more practicable; consequently the weak state of the limbs with the other symptoms of general disease sooner disappears."

The tendency of modern views, however, is to the disregard of this indication for rest, even in the earliest period.

Thus, Mr. William Adams, employs (Lectures on the Pa-

thology and Treatment of Lateral and other forms of Curvature of the Spine. By William Adams, F.R.C.S. Churchill,
London, 1865), very emphatic language :

"Since my connection with the Orthopedic Hospital, now
more than twelve years, I have invariably opposed this rule"
(of the horizontal posture) " and adopted the plan of applying mechanical support to the spine by means either of a
leather or steel apparatus, according to the age, and allowing
the patient to walk about when so disposed.

"This treatment I find to be applicable to all stages of
caries or destructive disease of the spine, from its commencement, which we may in most cases diagnose before any angular
projection has taken place, and even when the destructive
process is evidently advancing, provided the patient is able
to bear the support and is able to walk."

In spite of this high authority, the old authors were right
in principle, though they may have persisted in applying it
while the constitution needed air and exercise more than the
local disease needed rest.

These two necessities are the antagonists which divide the
practice, sometimes one and sometimes the other claiming
more than its share of attention.

5. Artificial support of the diseased vertebræ, taking off the
pressure upon the bodies, and obviating the strain upon the
muscles and ligaments attached to the spinous processes and
arches of the vertebræ, is seen at a glance to be theoretically
correct, although the measure may lead to abuse, by too much
trust in the supporting power of the mechanism, or trust in it,
while the utmost rest, and perhaps cold, are necessary to arrest
the destructive inflammatory process.

So far as the disintegration of bone and intervertebral cartilage may depend upon pressure and motion of vertebræ upon
each other, the annihilation of pressure and motion become
important elements of treatment. These therapeutic indications have been appreciated a much longer time than is generally supposed.

To illustrate the history of this subject these figures (Fig.

52) are introduced. They are taken from Shelldrake's " Treatise on Diseased and Distorted Spine," London, 1816, in which Shelldrake says: " In 1782, 1 published a description of this instrument with plate, and such observations as then occurred to me on the subject."

FIG. 52

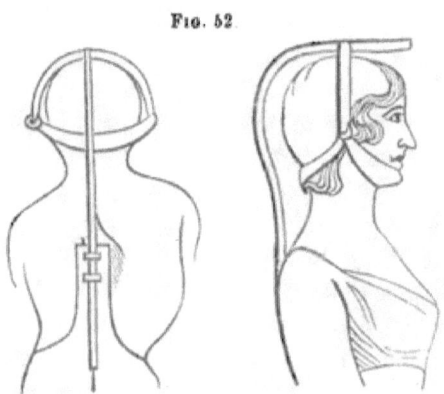

Shelldrake's apparatus for Pott's Disease, published 1782.

In Sir James Earl's Observations on the cure of Curved Spine, 1803, quoted in Shelldrake, this language occurs:

" If a machine be contrived to elevate the head and support the thorax, passing down the spine, and strengthening it *as a splint does a broken limb*, resting on the pelvis at its basis, and with a contrivance to give such gradual and permanent extension as the weak parts will bear without injury, and to be continued till, by a deposition of osseous matter, the yielding vertebræ become firm and compact bones, I am clearly of opinion that much good from it may be derived. . . . . . . What I principally wish is to show that it is safe and useful, and to set aside the disinclination which I perceive in many practitioners, as well as in the writings of Mr. Pott, to admit of its use or assistance in cases of curvature attended with caries; and further, I shall endeavor to make it apparent that in some of these cases, such a contrivance is not only frequently useful, but often absolutely necessary."

During this period, and until very lately, the plan of treatment of spinal caries by issues, advocated by Pott, held such sway that the importance of mechanical support obtained no general recognition in the views of medical men.

The present favor in which the treatment of vertical curvature of the spine by mechanical support is held, is very much owing to the experience of Dr. Henry G. Davis, of New York, and the restatement of the indications in a more philosophical manner. This expedient of treatment is founded upon the law of the disease to affect the bodies of the vertebræ, and to leave the arches, articulating surfaces, and processes untouched. The principle of treatment is that of a splint applied to the back of the spine to keep it from bending, and to force the weight of the parts above upon the articulating processes.

In the No. of the Boston Medical and Surgical Journal for August 4th, 1852, Dr. Davis, at the close of an article on lateral curvature, refers to an apparatus for Pott's disease, having then employed it for three years. In the American Medical Monthly for 1856, vol. 5, p. 212, &c., Dr. D., in speaking of the difficulties in adapting remedies for Pott's disease, remarks : "The common mode of constructing apparatus to sustain the weight of the body upon crutches is utterly useless. . . . . . The bodies (of the vertebræ) and the oblique processes afford the only perpendicular support. The distortion is produced by the removal of the bodies of the vertebræ by ulceration. As the line of perpendicular support falls between the bodies and the articulations of the oblique processes, the weight of the trunk above approximates the bodies of the two adjoining vertebræ, as the diseased one is removed by absorption ; the oblique processes now sustaining the weight of the trunk act as fulcrums, upon which the vertebræ are tilted or rotated ; thus, the spinous processes above and below are separated from that of the diseased vertebræ, the articulations of the oblique processes being the centres of motion. It is this form of the vertebræ which enables us to make use of the whole vertebral column as a lever to restore it. By apparatus we are enabled to throw the entire weight of the superincumbent body upon the oblique processes, and at the same time separate the bodies of the vertebræ adjoining the diseased one from it, the contact of which is constantly irritating and producing absorption. . . . . . . The same principle of treatment, viz., the separating of the diseased surfaces, and removing from them all irritation from pressure, is equally applicable to diseases of the hip-joint."

These quotations are made to this extent not only to vindicate American Surgery against foreign claims, but to secure the honor of originality to whom it rightfully belongs among Americans.

In the Transactions of the New York State Medical Society for 1853, is an article by Dr. C. F. Taylor, of New York, upon " The Mechanical Treatment of Angular Curvature," &c., in

Fig. 53. Fig. 54.

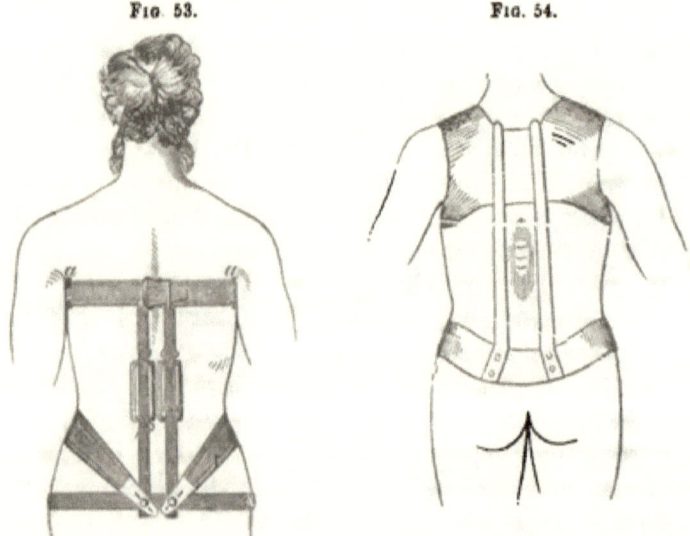

Apparatus of Dr. H. G. Davis for vertical curvature, arranged for disease of the lower dorsal, or upper lumbar vertebræ.

Instrument of Dr. E. Andrews for backward curvature.

which Dr. Taylor claims that he is the inventor of the only method which does not rely upon crutches to support the body in Pott's disease. From a letter from Dr. Taylor himself, we learn that his own apparatus was invented in 1859. This is several years after Dr. Davis had employed his apparatus.

The quotations already made show that this is not a modern invention, however original the conception may have seemed, and that it is one of the signs of that general change of therapeutics which has brought us to study more how to support the efforts of nature than to correct her deviations.

The apparatus of Dr. Davis, arranged for disease of the lower dorsal or first lumbar vertebræ, will be readily understood from the cut, Fig. 53.

Dr. Louis Bauer, of Brooklyn, has constructed a modification of this apparatus, which he calls a cuirass. It is made of woven iron wire, by fashioning it upon a plaster cast, and encircling its borders with an iron rod or wire, of sufficient firmness to resist any force likely to be applied to it. The fastening above, is by bands around the shoulders, and below, by a band across the hypogastrium. On either side are leather handles, for the purpose of carrying a small patient invested with this covering, as in transporting a turtle on his back.

The apparatus is made like the "wire breeches" of the same inventor, and when made a good fit it must answer the purpose admirably.

Figs. 54, 55, illustrate the apparatus adopted by Dr. E. Andrews, of Chicago.

The pelvic bands are made of metal properly padded and accurately shaped to the form so as to secure some *lifting* support to the parts above.

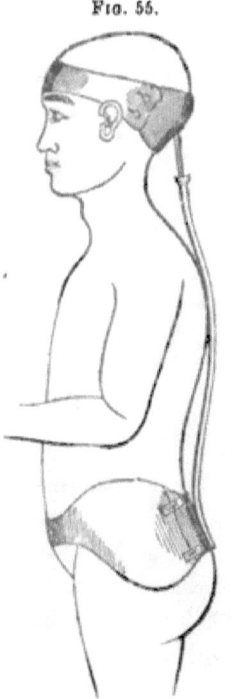

Fig. 55.

Head Extension for high Curvatures.

Three figures from Bonnet are introduced to show different modes of executing the same idea.

Fig. 58 shows what use may be made of a support like that

of Fig. 57. The patient, if a child, may be carried as if it
were in a basket, and if an adult he may swing himself and
thus secure relief from the tedium of entire helplessness.

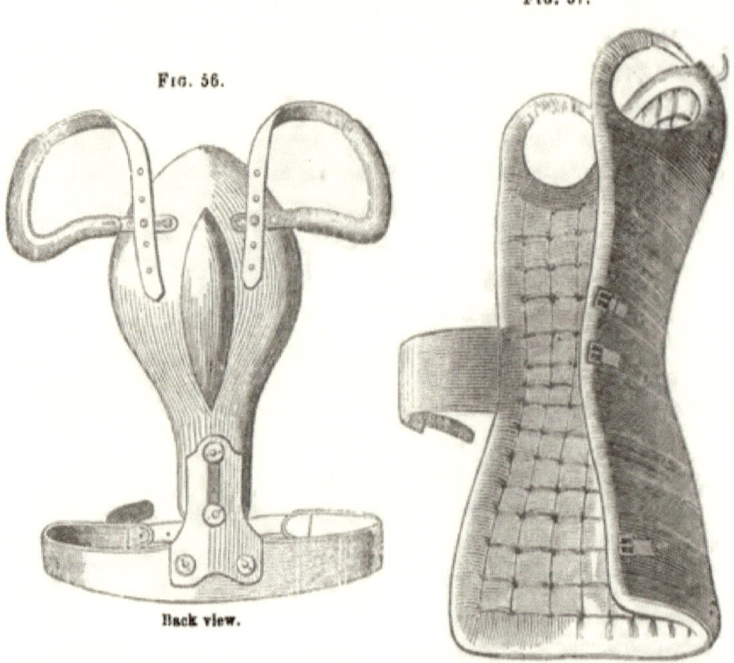

Fig. 57.

Fig. 56.

Back view.

There is in the apparatus devised by Dr. Taylor already
referred to, an arrangement of hinges by which the degree
of pressure upon the projecting portion of the spine may be
regulated by screws while upon the patient, which adds very
much to the adaptability of the apparatus. These hinges are
not intended for motion while the apparatus is worn, but only
for adaptation.

It is a desideratum, to have a plan which can be executed
anywhere away from mechanics especially skilled in the making
of such appliances. To this end, let a strip of hoop-iron be
cut, of sufficient length to extend along one side of the spine,
and another like it, to extend parallel with it along the other

side, to be separated, so that the spinous processes shall be free from pressure. Let the lower ends of these strips be riveted to a wider strip, which may be made of tin, to encircle one-half the body, passing across the upper part of the sacrum and iliac bones. A broad strap to fasten with a buckle, is to invest this piece of tin and encircle the body.

Fig. 58.

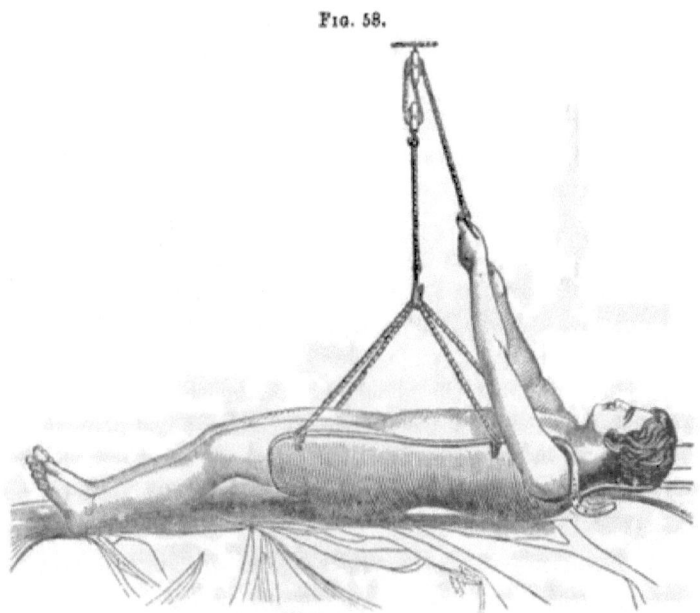

Bonnet's Spinal Shield arranged for swinging.

The lower ends of the vertical strips are thus held fast to the pelvis. The upper ends are to have a short cross-piece riveted to them, with an attachment at each end, for a padded strap going round the shoulder, with a short elastic at its fastening to the splint, to relieve the shoulder of the discomfort of an unyielding constraint, as employed by Drs. Taylor and Bauer, or the strips may be originally cut longer, so as to be bent in the form of a hook, passing over the clavicles in front. To give breadth of application to the ribs, let a piece of tin or sheet-iron of sufficient size, one for each side, receive two trans-

10

verse slits, through which the vertical strip is passed, before
it is riveted at both ends. The proper form of this metallic
splint may be secured by bending it across the knee, and ample
padding should shield thinly covered projecting bones from
pressure.

The following is a plan for an iron framework for apparatus
for vertical or angular curvature (Pott's Disease).

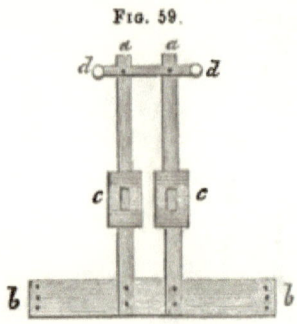

Fig. 59.

a a. Two parallel uprights. *b b.*
Base to apply to the posterior
half of the pelvis. *c c.* Plates of
sheet-iron, or thick tin, to be
padded, to apply on either side
of the affected portion of the
spine. They can be made to
slide upon the uprights till they
come to the proper place. *d d.*
Upper cross-piece, to each end of
which is attached a loop, well
padded, to encircle the shoulder. As neither elasticity nor
any other motion is desirable, the uprights should be made of
soft iron, in order to be easily adapted to the shape of the
trunk, and of sufficient bulk not to yield to the tendency of
the trunk to bend forward.

If the disease is in the upper dorsal vertebræ, it may be
better to make the uprights *a a* long at the top, to bend over
in front of the clavicles, in the form of hooks.

If the disease is in the sacrum or lumbar vertebræ, it is
doubtful whether any expedient but the horizontal posture will
obviate deformity.

It is believed that, for efficiency, a homemade apparatus
fashioned in this manner, may be equal to a more costly one.
To bring the means of treatment within the reach of all, is the
desideratum.

For disease of the lower lumbar vertebræ, this apparatus is
inapplicable, and one that directly lifts, can do but little good,
on account of the impossibility of avoiding the disposition of
the spine to bend forward. The difficulty is to extend the

lower end of the lever sufficiently below the diseased bones, to prevent the inclination of the body to bend forward. Anchylosis of the hip-joints would afford the desired condition. A mere lift acts upon the bodies of the lumbar vertebræ at too great mechanical disadvantage. Nothing remains but the horizontal posture.

For disease of the cervical vertebræ, the upper end of the apparatus must take hold of the head. The most feasible plan for this purpose is to employ strips of iron, as elsewhere explained for lateral curvature (Fig. 32 and 47), crossing them between the shoulders, like shears, so that a cross-piece may be fastened to their upper extremities, passing under the mastoid processes and lower jaw, to support the head and face. The constraint of movement may be very undesirable, but, for a patient that cannot otherwise hold his head up, it may be a very acceptable relief, as well as an efficient therapeutic agent. In this case, it is practicable to rest the weight of the head partly or entirely upon the shoulders of the patient, rendering it unnecessary that the apparatus should extend farther down than is required to give firmness of support to the splint employed as a lever.

In cases in which motion of the cervical vertebræ is not injurious, Shelldrake's apparatus, (Fig. 52), or some modification of it, may be easily constructed.

The cut from Ambrose Paré (Fig. 60), is of no little interest, showing the progress of ideas in three hundred years.

The notion seems to have been, that the spine should be relieved of its distortion very much as a dislocated shoulder may be pulled into place.

Ambrose- Paré describes a case which was probably one of Pott's disease. "I have lately," he says, "met with a case which appears like a dislocation of two of the dorsal vertebræ. On passing the fingers down the spine, there is a depression where the spinous processes of the 5th and 6th dorsal vertebræ ought to be ; the spinous processes are not to be felt.

"The man tells me that about a year since he was running and leaping with considerable violence, when he felt something

snap in this part of his back, where he has ever since had more
or less pain, with numbness of his lower extremities. He has

FIG. 60.

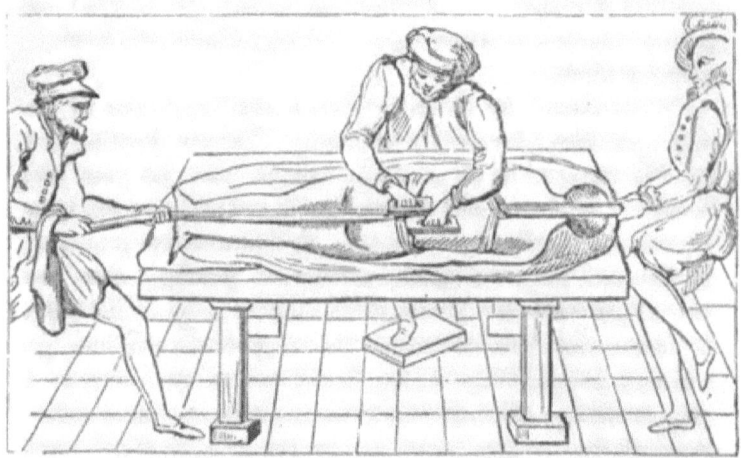

not since the accident been able to take much exercise." The
depression may have been the result of a compensating curve
(Clordosis) to compensate for a projection immediately below.
The projection, if not sharp, might in this case be mistaken for
the normal position.

6. Finally, the question arises as to the time, and the degree
of exercise, and of food, whether the good effects of exercise
on the general health will balance the local irritation from the
unavoidable jar and friction of the diseased parts. The question is nearly answered by what has already been said. In the
acute stage, and while the abortion of the disease is possible,
quiet is desirable for its influence on the general circulation, as
well as the avoidance of local irritation. Later; exercise
would be desirable for the general health, while it must be
destructive, by aggravating the local disease, and through this
irritation, increasing the general irritative fever. The extent
to which the spine may be supported by apparatus, will then
determine the propriety of exercise in the erect posture. Still
later, when **the** destructive processes have ceased, and the

reparative processes have commenced, there is great need of the invigoration of exercise and open air, with the amusement of the senses by changing scenes; but, without adequate support **to the** spine, there **is** the greatest danger of breaking down the weakened column, and exciting the inflammation afresh, or producing a palsy, by the pressure of the fragments upon the spinal cord.   Here, no little value of a splint is the safety afforded, while the general health is invigorated by exercise in the erect posture and in the open air.   Without this protection, the exercise, indoors or out, should be limited to such movements as can be executed in the horizontal posture, until, by cautious trials, it is found that the patient can endure shocks **without** uneasiness, and refrains from supporting the **trunk by** placing his hands upon his thighs.

The question of food is more simple, because the exercise of the gastric function produces no jar or friction of the parts diseased.   During febrile excitement, food must be abstained from, simply because it cannot be retained and assimilated, as in a similar general state from any other local disease, but, as soon as the effects of eliminating medicine, rest, and the recuperative tendency, render digestion possible, the introduction of nourishment is of the greatest importance, for a well-supplied nutrition is one of the most effective regulators of movements of the blood, perturbed by the unbalancing influences of inanition. Thus, a judicious succession or alternation of medication and diet may do what neither could accomplish alone.   The power of the digestive organs must always **be** regarded, so that hardly any other rules can be given, than the regularity of meals and the avoidance of too long fasting.   To this end, a cracker or a piece of bread at bedtime, may conserve the powers, and render the breakfast more easily digested, than would be the case after twelve hours fasting.

The temporary use of alcoholic stimulants, to excite the gastric glands by direct stimulus, or through **the** excitement of the brain, may often be extremely useful, but the permanent employment of any of their forms, as elements of diet, could not be productive of **any but pernicious** effects.

7. *Atmosphere.* It has been a desideratum for this, and for all other forms of scrofulous inflammations and degenerations, to secure an invigorating atmosphere, especially during the periods of necessary confinement.

It has only been within the means of the rich to carry the patients to **regions** of bracing and pure air, there to remain during the depressing heat of summer.

A recent invention of Mr. **A. S. Lyman,** has rendered it practicable to secure a pure and cool atmosphere, wherever new lime, new charcoal, and ice can be obtained.

The expedient consists in constructing a cupboard, which may be the head-piece of a bed, in which a sufficient amount of air entering at the bottom passes, first over new lime, by which it loses some of its moisture, its carbonic acid, and some of its organic impurities, and is warmed so as to acquire a tendency to rise, after which, in ascending it passes through charcoal, losing most of its remaining impurities ; then, turning at the top of the inclosure to pass down, it **flows among** cakes of **ice,** by which **it** is cooled and made heavier, so that it descends **with** accelerated speed, **to orifices above the bed, over** which it flows, and where it is retained by high sides or by curtains.

This makes it possible to give a patient confined in a horizontal, posture a better atmosphere than he could enjoy if he were moving about, especially during the warm season, and in places infected with malaria, or in cities whose atmosphere is charged with the effluvia of animal decomposition.

The tonic effect **of a purified** atmosphere for spinal and joint ulcerations, and **for most chronic diseases** requiring confinement, is likely to **remove the antagonism** between indications **for local** rest and for general exercise, and to secure to the local disease its necessary rest, while the general health gains **by pure air.**

TALIPES.

Definition and Classification of the Genus, Species, and Varieties of Talipes.—The term TALIPES (Latin, *talus*, an ankle, and *pes* a foot) has come to be adopted as a generic term for what is known as club-foot, reel-foot, and splay-foot, or flat-foot. The name expresses only a minor element of the deformity,—the ankle, in some species, being not at all displaced or deformed; but this is of no great importance, since the technical signification has been agreed upon.

Definition.—A malposition or malformation of the foot, congenital or acquired, in which, from some deviation at the ankle-joint, or in a greater or less number of tarsal or tarso-metatarsal joints, the sole of the foot fails to apply to the ground in the natural position.

Of this genus there are six species in two groups:

	1st group.		2d group.	
1. Talipes Equinus,		4. Talipes Valgus,	
2. " Dorsalis,		5. " Plantaris,	
3. " Varus,		6. " Calcaneus.	

Of these species there are six possible secondary combinations or varieties, namely:

1. Talipes Equino-Dorsalis,	4. **Talipes Valgo**-Plantaris,
2. " Equino-Varus,	5. " " Calcaneus,
3. " Calcaneo-Varus,	6. " " Equinus,

with the possible existence of a T. Calcaneo-Dorsalis.

The conceptions of the tertiary combinations, when once familiar, will also be simplified by classifying them thus:

Talipes Equino-**Varo-Dorsalis**, Talipes **Calcaneo-Varo-Dorsalis**,
    "   "   Valgo-Plantaris, "   "   Valgo-Plantaris.

Talipes equinus is the term applied to that position which, by long-continued voluntary elevation of the heel to compensate for several inches shortening of the limb, becomes not only habitual, but fixed by the permanent shortening of the triceps extensor pedis, and the adaptation of the ligaments to the habitual relations of the bones of the leg and tarsus. The habitual voluntary contraction of the triceps muscle, gastrocnemii, plantaris longus, and soleus, terminating in the tendo Achillis, becomes permanent and involuntary; after which the muscular tissue changes its character; is absorbed or in part replaced by fat, while the white fibrous tissue investments become hypertrophied, converting the muscles into ligaments both in constitution and function. The result is a compensating deformity; and in order to attain the best possible compensation, bringing the phalanges as nearly as possible within the vertical line of pressure, the foot comes to be more than naturally arched by the contraction of the tibialis posticus, the peroneus longus, and the flexor longus digitorum, upon the back of the leg; and by the adductor pollicis, the flexor brevis digitorum, the abductor minimi digiti, and the musculus accessorius, with corresponding shortening of the plantar fascia under the foot. The action of the long and the short flexors of the toes, would curl them under the sole, as the fingers are flexed upon the palm, if they were not kept out by the weight of the body upon the phalanges.

This makes the variety T. equino-dorsalis, which, in the confirmed state, is more common than either species unmixed. The deformity which has been described as originating in a voluntary attempt at compensation, may result from spasmodic contraction of one set of muscles, or paralysis of their antagonist.

Fig. 61.

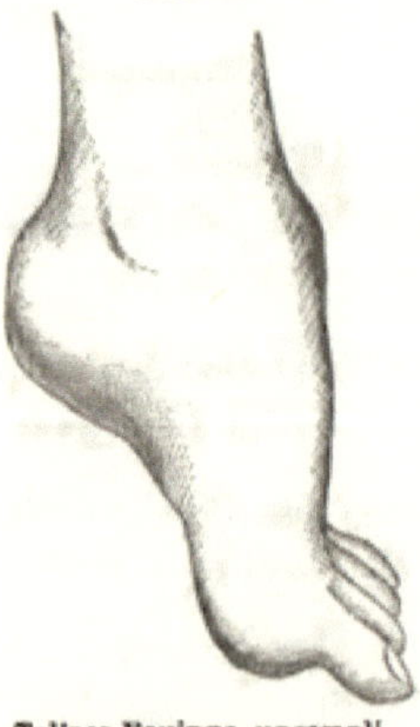

Talipes Equinus, uncomplicated.

Fig. 61, illustrates Talipes equinus, without a very marked arching of the foot; the heel is simply elevated, but a persistence of the malposition and the weight of the body upon the foot invariably arches it, producing T. equino-dorsalis.

The next figure, from Bigg (Fig. 62), exhibits an extreme Talipes dorsalis (T. cavus of Barwell), in which the calcaneum and metatarsal bones are brought nearer, by the

contraction of the plantar muscles and the permanent shortening of the plantar fascia.

A good representation of the deformity should show less
abruptness forward of the heel, the arch corresponding with
the articulation between the astragalus and calcaneum behind, and the scaphoid and cuboid bones before.

The shape of the foot produced by the Chinese shoe, is a
shortening of its length and a humping up of the instep, making
a stumped appearance, a talipes dorsalis.

Fig. 62.   Fig. 63.

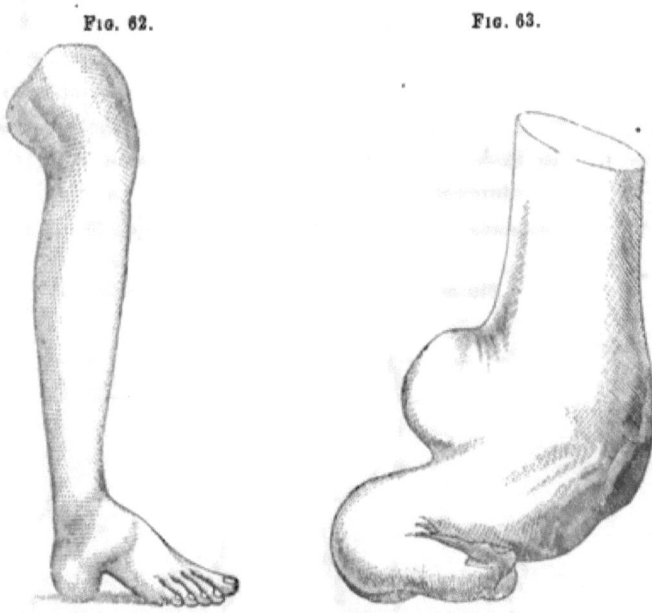

Talipes Dorsalis, extreme.   Talipes Equino-Dorsalis (from Detmold).

A modification of this species may also result from a partial dislocation by violence, breaking up the ligamentous fastenings on the dorsum of the foot, and permitting a shortening
of the base of the tarso-metatarsal arch. This once occurred
under the observation of the writer,—a young man falling
twenty feet from a tree, and dislocating the tarso-metatarsal
articulations of both feet. The deformity was never completely reduced, and the tarso-metatarsal joints remained

permanently elevated, requiring shoes to be made according to special measurements.

The combination of these two elements with the extension of the toes, occasioned by walking upon this plantar aspect, is shown in an extreme degree in the preceding figure (Fig. 63), which is taken from Detmold.

There is also here another element, that of inversion, which makes it to some extent a T. varus, but not to a sufficient degree to call it a T. equino-varus.

Talipes Varus.—This is the most common of all the species, and consists in the inversion and rotation of the anterior half of the tarsus, which can, to a slight degree, be imitated by taking hold of the phalanges and metatarsus, and bending the foot in the direction in which the tibialis anticus would draw it. In making this twist, the calcaneum and astragalus will become adducted, as in the position which a child will sometimes assume, in standing upon the outer edge of the foot.

Fig. 64.   Fig. 65.

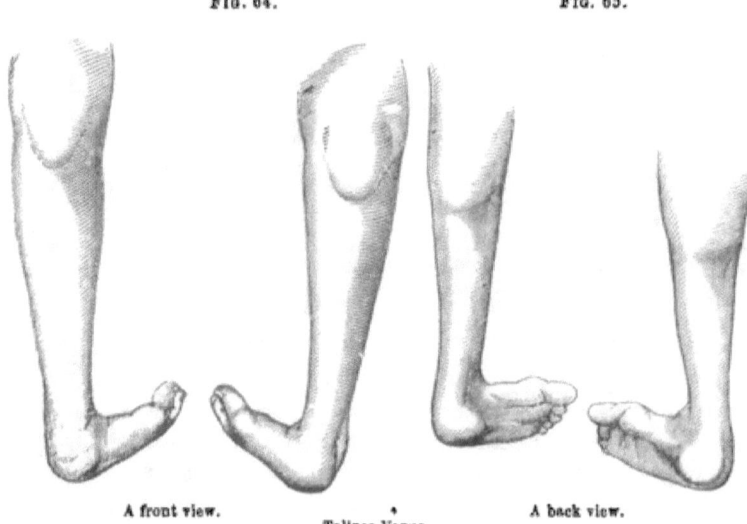

A front view.   A back view.
Talipes Varus.

This malposition is very well illustrated by the preceding cuts (Figs. 64, 65), representing the lower extremities of a

gentleman fifty-two **years of** age, whose parents took him to
Cincinnati, when an infant, to consult the best surgeons of
that city.   The parents were told that nothing could be done
for the child.

Attention has been called to a better anatomy of this deformity, by Mr Barwell, in his little book, entitled, " Club-Foot
without Division of Tendons," in which he gives the appropriate
name "medio-tarsal articulation," to the articulation between
the calcaneum and the cuboid on the outside, and between the
astragalus and the scaphoid upon the inside.   " **This is the cen-**
tre of the twist, which, in a delicate foot, can almost be imitated
inward, while **outward,** or in the opposite direction, there is
very **little capability of a twist to bring down the inner side
of the sole."**

In this species there is no important contraction of the triceps through the tendo achillis, or, in other words, a corresponding elevation of the heel.   The heel is tilted over as if
the hand were adducting the whole foot, by taking hold **of the**
foot and pulling it inward.   The inner or tibial edge of the
foot is turned up, and the outer or fibular side turned down,
and, in the worst cases, carried in toward the opposite foot, so
that the outer side of the dorsum of the foot comes to the
ground.   The sliding of **the scaphoid outward upon the** astragalus, makes the former bone very prominent, receiving, with
the cuboid and the anterior portion of the outer and lower
edge of **the calcaneum, the** weight **of the body,.in** standing
and **walking.   The cuticle** becomes **unnaturally** thickened,
and between **the integument and the bones, bursæ develop**
themselves **as cushions to protect the bones from pressure in**
walking.   ·

There is at first no transverse narrowing of the metatarsus
and phalanges, but the pressure of walking gradually approximates the two borders of the metatarsus and phalanges ; the
fissure or concavity being in the plantar surface.   The deformity appears to result from disproportionate contraction of
the tibialis anticus, while the flexors and extensors are balanced, and the peronei muscles paralyzed.   **The tibialis pos-**

ticus assists in the inversion of the foot, so as to make the toes point toward the opposite foot.

Talipes Equino-Varus.—This combination is the most common variety of talipes acquired subsequently to birth, and consists of disproportionate contraction of the triceps extensor pedis through the tendo achillis, elevating the heel and making a talipes equinus. The tibialis posticus tends to double the foot inward, while the tibialis anticus, at the same time, acts upon the inner edge of the foot, and elevates and rotates it, while the tibialis posticus, flexor longus digitorum, and the short flexors originating from the calcaneum, shorten the arch of the foot, making the compound expressed by the succession of terms, talipes equino-varo-dorsalis. Walking doubles the foot still more, antero-posteriorly as well as transversely, almost completely turning it upside down, giving the gait a much worse hobble than that of simple varus, and presenting a complicated deformity, requiring apparatus equal to the versatility of the hand for its successful treatment.

It has been neglected to illustrate this by a good figure, but its elements must be sufficiently apparent from the descriptions already given. (See Fig. 63.)

These three species constitute a natural group, making all degrees of combination ; and the next group is equally natural, though its species are of less frequent occurrence.

Talipes Valgus.—The condition in which the anterior half of the foot is carried outward in the direction opposite to that of T. varus. The tibialis anticus and tibialis posticus fail, and the peroneus longus and P. brevis, passing behind the external malleolus, pull upon the outer side of the foot and evert it. At the same time the peroneus tertius passing down in front of the external malleolus, elevates the outer side of the foot, and tilts the astragalus and calcaneum outward, in the opposite direction to that taken in T. varus.

The following cuts (Figs. 66 and 67) illustrate this species, which is rarely met with without complication.

The figure (66) is taken from the cast of the foot of a gentleman living in Boston. The cast is kept by Messrs. Tie-

mann & Co., surgical instrument makers, New York, for the
purpose of making upon it the apparatus which aids him in
walking.

Fig. 66.   Fig. 67.

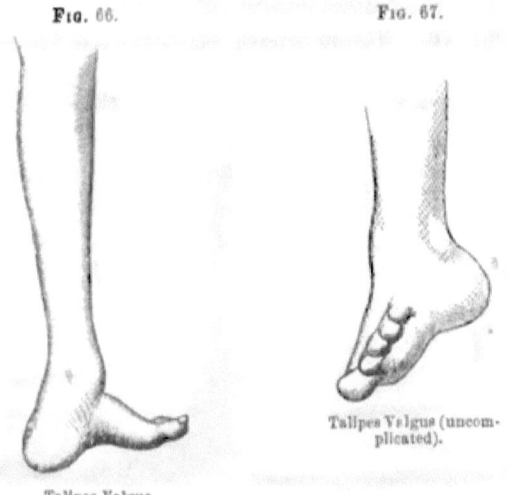

Talipes Valgus (uncom-
plicated).

Talipes Valgus.

The figure is seen from behind and on the inner side.

It will be noticed that this is a simple T. valgus, without any
flattening of the arch of the foot, to make the species plantaris.
A better illustration of T. valgus is seen in the annexed figure (Fig. 67).

The preponderating action of the three peronei muscles
elevates the outer side of the foot. The foot in walking is
brought into a position which brings a great strain upon the
plantar fascia, finally resulting in its elongation with the production of T. plantaris.

TALIPES PLANTARIS.

A minor degree of T. plantaris is simple flat-foot (Fig. 68),
which is often congenital in the Caucasian race, and is almost
universal in the African.

The next figure (Fig. 69) illustrates T. plantaris in a
marked degree, the extensors of the toes acting so as to re-
verse the arch of the foot and cause the " hollow of the foot to
make a hole in the ground."

The same perversion of muscular contraction has only to proceed to a greater extreme in force and duration, and the phalanges become entirely lifted from the ground, so that the weight of the body comes altogether upon the tarsal bones, as seen in Fig. 70. This deformity has been called *talus*.

FIG. 68. FIG. 69.

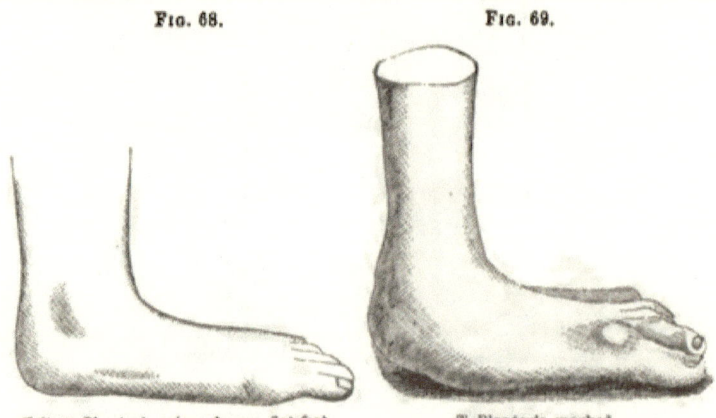

Talipes Plantaris, minor degree, flat-foot.   T. Plantaris, marked.

The combination of the two species just described often occurs, making **T.** valgo-plantaris, **generally** produced by perverted innervation during growth.

FIG. 70. FIG. 71.

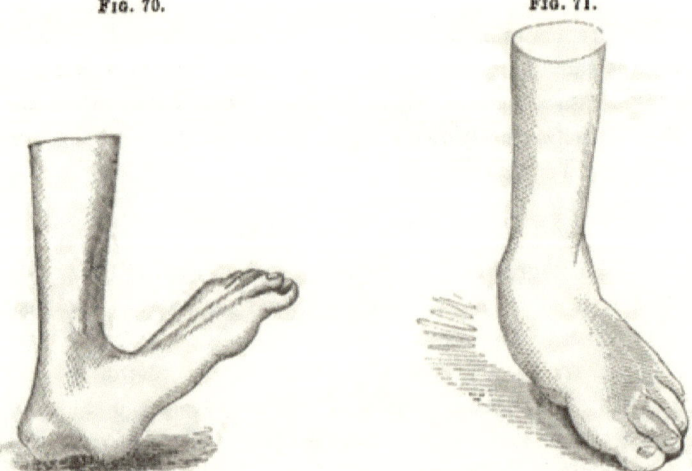

T. Plantaris, extreme (from Scoutetten).   T. Valgo equinus (from Tamplin).

The unresisted contraction of Achilles' tendon, in connection with that of the peronei muscles, makes a T. valgo-equinus, as seen in Fig 71, or a T. equino-valgus when the elevation of the heel is the preponderating element.

Talipes Calcaneus.—A deformity in which the heel comes to the ground, and the anterior portion of the foot is drawn up by the disproportionate contraction of the tibialis anticus, peroneus tertius, and extensor longus digitorum.

This is a congenital deformity well shown in Fig. 72 (from Detmold).

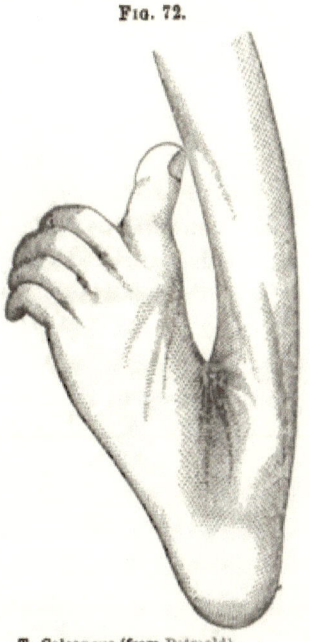

Fig. 72.

Barwell says that this species must be extremely rare; for though he has had large opportunities of observation, he has never seen a case. But one case has occurred to the writer, and that in a new-born infant, the dorsum of the foot lying against the tibia. Some degree of valgus or outward deviation of the ankle-joint complicated the case.

In this instance, the muscles of the leg were all well developed, so that the deformity could not have occurred from paralysis.

The deformity was more satisfactorily explained by the supposition of constraint of the foot

T. Calcaneus (from Detmold).

by position within the uterus. The deformity readily yielded to moderate persistent force.

All the possible degrees and combinations will be at once comprehended by familiarity with the classification, aided by the illustrations.

Similar deviations from the normal form of the hand should receive a similar classification, only that their rareness makes it unnecessary. Their pathology is doubtless the same, whether con- or post-genital, depending upon paralysis of one class of muscles, or overaction of their antagonists, or both combined ; or more rarely, some accidental injury, resulting in partial dislocations ending in permanent deformity, or from the contraction of the cicatrices of burns or ulcers.

1. *Complications.*—The complications may be congenital or acquired ; absence or diminution of one or more bones, implying the impossibility of complete restoration of the form and functions of the foot, though great improvement may, in some cases, be effected by treatment.

2. Anchylosis of one or more joints from fractures or wounds, nearly or quite hopeless of benefit from subsequent treatment.

3. Anchylosis from arthritic or periosteal inflammation, in which the treatment is chiefly preventive, by substituting, before it is too late, passive motion for absolute rest of the parts in relation to each other.

4. Contraction of cutaneous cicatrices from burns, ulcers, or wounds. The treatment should be preventive, for confirmed deformity, from these sources, is extremely difficult to overcome.

5. Rheumatism, producing talipes, or simply attacking a taliped, requiring the abatement of the rheumatism in addition to whatever else may be done.

6. Corns and bunions requiring nice adaptation of shoes where, from the age of the patient, they cannot be cured by restoring the foot to its proper form.

7. Absence or deficiency of toes.

°8. Supernumerary toes which may be cut off.

9. Deviation of the forms and directions of the toes from fractures, wounds, arthritic, or periosteal inflammation, the contractions of cicatrices from burns or other injuries, from faulty shoes, from pressure of the weight of the body, or from paralysis of muscles. These deviations are sometimes incapable of remedy except by amputation of the offending toes.

Causes and Nature of Talipes and Allied Deformities.—The nice adjustment of forces, by which typical symmetry is produced and maintained in all organized growth, only needs to be contemplated to secure admiration.

The exceptional deformities proving the possibility of imperfect adjustment of these forces, or of the occurrence of accidental impediments to their exercise, only excite our attention all the more, to the nice balance observed in the ordinary working of the law of development.

In individual failures of this organic law of symmetry, the question will arise as to the modes of deviation:

1. Whether from **excessive** nutrition, analogous to that which secures the disproportionate growth in parts which are brought to perform compensating functions, as a **leg or a** kidney, which from the impairment or destruction of the opposite, is invited to perform more than **its natural part.**

2. From deficient nutrition direct, **from the** obstruction **of** the bloodvessels **which** supply **it, or** indirect, from failure of **nervous supply** to the capillaries **of** a part, failing to open them to the supply of blood, or from accidental or artificial quietude, analogous to that of muscles closely confined in splints and bandages, while a fractured bone is uniting.

3. From accidental positions, widely varying from those which are usual and which act to produce deformities, like the forces which are afterwards employed to remove them. By this means, some tendons may be forced to grow too long, and others permitted to become too much shortened, while the bones which become inordinately compressed take the shapes which the altered forces tend to give them.

4. From some observations made by Cruveilhier, this careful pathological anatomist came to the conclusion that position of the foot, within the uterus, was often the cause of talipes.

As a moderate talipes varus is the ordinary position of the foot within the uterus, this deformity can hardly be explained upon this hypothesis; but a talipes valgus might possibly be produced by an eversion of the foot from the pull of the umbilical cord accidentally entangled around it.

Twisting and displacements and spontaneous dislocations of the knee-joint, of the hip-joint, and of the shoulder-joint, can sometimes be most plausibly explained upon this supposition.

5. From the occurrence of causes which directly compress, or

11

partially or completely cut off, portions of the developing limbs; portions of the liquor amnii unusually condensed or solidified into sheets or shreds, may **produce** deep fissures in parts upon which they press; or they may completely amputate the included **parts.** The peculiar deformities constituting the genus **talipes** can hardly be explained by reference to this class of **causes.** Spontaneous amputations **doubtless** often owe their occurrence to this cause.

6. From disease directly resulting in the death of **the parts** affected. **The writer** has in his possession an aborted **fœtus** of four months, which exhibits gangrene of one upper extremity, including the shoulder. If this fœtus had lived, there **would** have been the birth of a one-armed child. Spontaneous **amputations** are sometimes produced by this cause, but talipes cannot be thus explained.

7. From the union of parts of two or more individuals, resulting in redundancy of number. **This is the** explanation **of a great variety of** monstrosities, but it does not apply to **talipes.**

8. From an influence existing in the germinal origin of the individual, like that which determines the color of the skin, the family likeness of features, and the temperament. It is thus, that in some families there is a perpetuation, through several generations, of five fingers upon the hand and six toes upon the **foot, the deficiency of** a thumb or a redundant one.

Though several **cases of talipes sometimes occur in one** family, and in rare cases it may be repeated in the next generation, the cases are too few to favor this explanation of its occurrence. Causes acting upon the innervation of the fœtus, subsequent to the formation of the type of the individual, constitute a more probable explanation.

9. From **causes set in operation** through physical and mental influences of the mother. As an example of physical influence, **one of the common expedients** for distinguishing pregnancy **from** enlargements within the abdomen from other causes, is to place the hand, previously reduced in temperature, upon the mother's abdomen, to excite a convulsive movement in the

fœtus. This movement may be simulated by the compression made by the sudden tension of the abdominal muscles induced by the cold application.

On the other hand, great physical exertion, and the occurrence of grave disease affecting the constitution of the circulating fluids, are followed by diminution or cessation of the fœtal movements, as if from some diminution of the fitness of the blood to afford to the fœtus the highest activity of nutrition. The death of the fœtus, and its expulsion is a frequent occurrence under these circumstances.

That deformities should sometimes arise from this impaired or perverted nutrition, is as probable, as that similar disturbances should, after birth, produce local congestion and inflammation, or convulsions and paralysis; some constitutional tendency, previously induced, determining the location and character of the diseased action.

Protracted mental depression, the indulgence of ungoverned anger, hate, or revenge, impairing digestion, are supposed to be unfavorable to the best development of the fœtus, while the cheerful and joyous emotions are invited as most favorable.

With the shock from the sight of a repulsive object, the mother feels a convulsive movement of the fœtus, followed by a diminution of the habitual movements, and her attention is afterwards anxiously fixed upon her own sensations and those produced in her by the fœtus.

Derangement of the digestion of the mother, and the consequent impairment of the healthy and nutritive qualities of her blood, which is the source of nutriment to the fœtus, often exist for a longer or shorter period, and deformities sometimes follow, but at the birth, the mother's fears are generally found to have been needless, as a perfect form occupies the place of the dreaded deformity.

In the few cases that do occur, there are, in exceptional instances, striking resemblances to some object seen by the mother during pregnancy; but upon close scrutiny of the deformities, they are found to belong to classes of excessive, deficient, perverted or arrested development already referred to, from the various causes classified ; and these resemblances are **too few,** in comparison with the whole number, to be worthy of any other explanation than that of coincidence. We all know how a striking coincidence takes more hold upon the mind than many discrepancies. The adoption, early in the civilization of all nations, of the theory of the direct production of special deformities through the images impressed upon the mind of the mother, is probably thus best **explained.**

The deformities arising from spasm and **paralysis are** more frequent in the lower extremities, from the more feeble, more easily deranged, and less easily restored innervation **of those** parts. They are, therefore, more often seen in the streets, **and** from the awkward movements in walking, they are more repulsive than deformities of the upper extremities, which need not **be** made conspicuous in public places.

The late **development** and comparatively low innervation of **the** inferior half of the fœtus, might be expected to result in the existence at birth of a greater number of deformities, produced by nervous derangement, in the inferior than in the superior half of the body. From this physiological order of development, as well as upon the hypothesis of coincidence, therefore, a mother who is shocked at the sight of a lame leg **is** more likely to have a child affected with talipes, than with a corresponding deformity of the hand, the detoriating influence **of the nervous** impression upon the blood being more likely to result in spasmodic or paralytic affections of the lower than of the upper extremities of the fœtus.

As the various species of congenital talipes are **similar to** the corresponding deformities developed subsequently to birth, from derangements of innervation, it is fair to infer, that in most cases, a similar derangement of innervation has existed during fœtal life. This conjecture is rendered more probable by dissection, which shows that the bones of the tarsus have their proper forms until they are afterward slightly changed in figure, by the great pressure to which they are subjected in walking. This change is much less than a superficial glance would lead one to suppose, there being nowhere a complete dislocation, but only a sliding a little further than the normal length of the ligaments permits.

The following figure (Fig. 73) taken from Little (On the Nature and Treatment of the Deformities of the Human Frame), sufficiently illustrates this point:

The relative importance of paralysis and spasm, in the production of this and other deformities, will be differently appreciated by different minds, standing in opposite positions. The following quotation from Bauer,* representing the older pathology, and from Barwell,† representing the newer, illustrate this point.

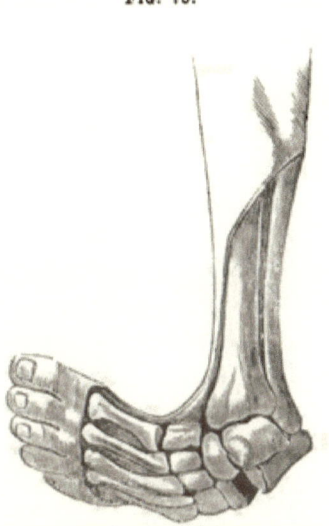

Fig. 73.

Dr. Bauer (p. 12) thinks contraction of the sural muscles (the muscles ending in the tendo Achillis), generally the chief cause of the extension of the foot in talipes equinus. He makes no account of the doubling up of the foot at the medio-tarsal articulation, so carefully explained by Little and Barwell, and, equally with Barwell, omits to mention the calcaneo-metatarsal and calcaneo-phalangeal muscles, as elements in the etiology.

Referring the disease to the shortened muscles, he says, " As a general thing, the contracted muscles have lost all susceptibility of being acted upon by the galvanic current, yet their powerful extension gives rise to unbearable pain. This fact seems to demonstrate that the muscular structure is in a

---

* Lectures on Orthopedic Surgery, by Louis Bauer, M.D. Lindsay & Blakiston, Philadelphia, 1864.

† The Treatment of Club-Foot without the Division of Tendons, by Mr. Richard Barwell, &c. London, 1863.

state of contraction to the extent of its capacity, or the substituted tissue is void of all contractile" (expansive) " power. It is certain that innervation has not been entirely lost, while pain can be provoked by extension."

In the conditions referred to in this paragraph, the occurrence of pain may, perhaps, be better explained by bearing in mind that the muscles concerned have, for the time, acquired the conditions of ligaments.

We know well enough, that ligaments are susceptible of acute pain when overstretched. When a muscle, therefore, which has lost its function from loss, change, or paralysis of its muscular substance, is pulled further than its investments of white fibrous tissue will permit, without violence to its habitual physical condition, it is in close analogy with an overstretched ligament, and it should be the seat of pain, the same as if it had originally been a ligament.

The following additional quotation is a further illustration of the spasmodic pathology :

" After the division of tendons, many months may elapse before the galvanic current makes any impression, and in some instances the contractibility of the muscles is gone forever."

If the division of tendons is all that is done, the shortening ought to go on still more. It is, probably, the subsequent movements, effected in the course of the treatment, that restore the susceptibility to the galvanic current.

The theory that permanent spasm is the uniform cause of distortions, finds an advocate in Dr. Joseph Pancoast, of Philadelphia, who claims that the elevation of the heel in talipes equinus is owing to the contraction of the soleus, while the gastrocnemius remains flaccid ; and he accordingly divides the soleus muscle by passing a knife in under the gastrocnemius, instead of the usual easy method of dividing the tendo Achillis.

It is found in any confirmed case of talipes equinus or T. equino varus, that the soleus is rigid and incapable of extension, while the gastrocnemius is yielding. Dr. Pancoast is, therefore, of opinion that the soleus is the author of the mischief. The fact has another explanation. When a muscle contracts with such power that its antagonists cannot extend it, the more powerful muscle soon becomes inextensible, and it settles into the function of a ligament, holding firmly the **points** to which it is attached, the muscular tissue gradually becoming atrophied, and while the size of the muscle diminishes, its hardness increases.

This is the state of the soleus in extreme talipes equinus. The upper end is attached **to the** tibia and fibula ; and when the calcaneum is elevated as far as its ligaments and bony connections will permit, the soleus can contract no **further, and if not** lengthened by an opposing power, it at length becomes hard and unyielding. This result is prevented in the gastrocnemius, **by its attachment to the femur,** whose movements keep this muscle active and extensible. After the soleus has become rigid **from** *immobility*, the gastrocnemius continues to have *mobility*, and, therefore, it preserves its extensibility. **It is** not that it draws less, but that it never acquires a stationary contraction, and, therefore, never comes into an unyielding condition.

Disproportionate weakness of the flexors of the foot, with anchylosis of the knee-joint, would probably result in equal extreme contraction, and consequent rigidity of gastrocnemius and soleus alike.

This explanation entirely destroys the value of Dr. Pancoast's method of dividing the soleus instead of dividing **the** tendo Achillis, in permanent elevation of the heel.

It would be wrong to leave the reader with the impression that Dr. Bauer considers spasm the uniform cause of **talipes,** and the following quotation, from p. 19, of his book, will do him justice in this respect :

" After mature deliberation, we have come to the conclusion, that the cause in congenital as well as acquired club-foot, is pre-eminently defective innervation ; and there is truly no reason why derangements in the nervous system should not take place in the fœtus as well as in the new-born child. In club-foot, the tibial nerve is the bearer of the difficulty, as must be inferred from the experiments of Bonnet."

. . . . . . "All forms of varus are caused by either muscular contraction or motor paralysis, and the individual bones of the foot yield only so much in their respective positions, as they are forced to do, by the abnormal muscular traction, and the superincumbent weight of the body. Being held for some time, and acted upon in the preternatural position, they gradually mould themselves accordingly, and become consequently malformed."

In the opposite pathological view, it is claimed by that careful observer, Mr. Richard Barwell, that it is not usually spasm of the stronger, but paralysis of the weaker muscles, which lies at the foundation of the deformity, and in support of this view he refers to the common experience, that in talipes the temperature is generally low, while in spasm it is generally high.

"Infants, as is well known, are subject to convulsions; and it is averred that sometimes one or more muscles, which have, during the attack, drawn the limb into malposture, do not recover from the contraction, but continue to keep the limb distorted. . . . The state should be one of persistent unvarying spasm, powerful enough to overcome the antagonistic healthy muscles, and permanent enough to produce lasting change of form. Such condition does not only never come under our notice, but it is, I believe, pathologically impossible. There are, no doubt, a few cases of peculiar paralysis of the voluntary power over the muscles, while the excito-motory function continues; and in the spasm of the whole set, the strongest organ will of course predominate. Voluntary power is as much used to control as to excite. The paralysis of this power is as much evidenced by violent and uncontrollable spasm, as by incapability of subordinate movement. In my experience, such state seldom continues long, unless there be cerebral disease or deficiency, but terminates, within a limited period, in death or complete recovery, or in simple paralysis in one set, and perfect restoration of power in another set of muscles."

. . . . . "Laryngismus stridulus, or false croup, is attributed, by some, to spasm of certain muscles; while by other authorities, and I believe with more reason, it is considered as paralysis of a different pair. Let it be observed, also, that the squint which may come and go with other symptoms of brain mischief, or may be a permanent affection, is certainly to be more rationally regarded as want of action in the outer rectus, which appropriates the whole of one nerve (the sixth), than as spasm of the inner rectus, whose nerve supplies four other muscles of the eye and appendages. Certain also it is, that some congenital deficiencies of the nervous system, whereof club-foot and club-hand are pretty constant accompaniments, as acephalosis, &c., &c., may, indeed must, produce paralysis, but there is no evident connection between such deformity and spasm." (P. 23.)

"Altogether, there can be no doubt that paralysis is very much more frequently the cause of club-foot than the opposite condition. . . . . The morbid contraction of a muscle, or set of muscles, is hardly ever violent enough, or persistent enough, to cause a permanent alteration in the shape of the foot, where the opposers remain active."

"The muscles, while healthy, are always kept at a certain degree of tension by tonic contraction, but when any one organ or set of organs has lost its power, the opposers draw the limb in the opposite direction, by virtue of that constant and elastic sort of force. For a long time after the commencement of the paralysis, there is nothing whatever wrong with the active muscles; but after they have been allowed to become thus short, for a certain period, they begin to adapt themselves to the shortened condition, and still further contracting, as they meet with no resistance, determine at last changes of form in other structures, and so produce permanent deformity." The clearness with which the points are here made, justifies the length of the quotations.*

Treatment.—It is believed that a careful consideration of the nature and pathology of the different varieties of talipes, as explained in the preceding pages, will afford the foundation for clear ideas of the indications of treatment, whether pre-

* See the discussion of spasm and palsy, p. 26, *et seq.*

ventive or curative. The plans and expedients for meeting these indications are now **the** earnest study of those interested in this branch of surgery. No words of mine can be more appropriate than those of Barwell. (P. 25.)

"**It is not** to be imagined, that when the limb has yielded **in the** direction of **the** healthy **muscles, the** sickly ones can **recover** sufficiently quickly or **entirely to** restore, by their unassisted might, the proper balance of **the foot.** The weakened muscles want assistance; and the way to render this, in the manner which shall best aid them to overcome **the** deformity, and to recover from the paralyzed or enfeebled **condi-**tion, is the problem which surgeons should endeavor to **solve.**"

It is one of the points showing the impossibility of **practi-**cally and completely separating Medicine from Surgery, and the different branches of Surgery from each other, that in these cases of paralysis, previous to the occurrence of obvious deformity, the disease would be said to be in the department of Medicine, **though mechanical or** chirurgical means are ne-**cessary to prevent the occurrence** of deformity; and after-**wards,** when the deformity places the disease fairly in the department of Surgery, the best period for surgical treatment **has been allowed to pass** by: because the case was in the department of Medicine.

The physician must study Surgery, and the surgeon must study Medicine.

Whoever has examined a case of club-foot, by taking hold **of it** with his hands, may have thought, that if he only had some machine that would take hold of the foot as firmly, and yet as tenderly as does the hand, without relinquishing its grasp, and yet pulling yieldingly but persistently and **without** tiring out, he could cure any case. The defect of every me-tallic apparatus is, that while it grasps the foot firmly enough, it pulls unyieldingly, without that distribution of force among all the distorted joints, which is effected by the hand. They are, most of them, intended to act chiefly upon the tibio-tarsal joint, while the most careless inspection of any species of talipes, **except** one of simple talipes equinus, will show that the dis-**tortion of this** joint is **a minor** element **in the case.**

That an adequate substitute for the hand is a desideratum not yet furnished to the public, is sufficiently proved by the words of Dr. Bauer. (P. 23.)

" There is no mechanical apparatus, however ingeniously constructed, which could be substituted for the hand, in the treatment of talipes, with any approximate degree of efficiency. In fact, if we could, without interruption, employ the hand as a mechanical agent, we should relieve most obstinate forms of talipes, which *too frequently* withstand our mechanical appliances." This is an estimate of the importance of some substitute for the hand, with an expression of hopelessness as to its attainment.

On the other hand, Dr. Gross, in his great work on Surgery, vol. ii, p. 1011, is well enough satisfied with our present attainments in the art, neither desiring nor expecting any improvements. He says, " It is perhaps not going too far to affirm that these topics" (club-foot) " are as well understood now as they ever will be."

Dr. Bauer again places this estimate upon our present attainments (p. 28): " They" (mechanical appliances) " possess no curative virtues, but retain the foot in the position in which tenotomy and the acting hand left it."

It is believed that, in the course of these pages, a process will be explained, which is a pretty adequate substitute for the hand.

The earlier experimenters in this art seem to have relied chiefly upon wood and iron, as substitutes for the hand; but so generally did they occasion ulcerations of prominent parts, that the art made no important progress until the introduction of subcutaneous section of tendons, by Stromeyer, in 1831. In a large proportion of the cases of talipes, including all the species equinus, the division of the tendo Achillis, permits an immediate improvement in the position of the foot, and facilitates the further reduction of the distortion of the joints of the tarsus. This tendon had been cut at various times before Stromeyer, by making an open wound; Isaac Mincius having divided it in 1685; Thellenius in 1784; Sartorius in 1806; Michaelis in 1809; Delpech in 1816; but none of these men could think of so simple an expedient as passing in a small knife at a point distant from the tendon, and so dividing it, that the incision through the skin should heal without suppuration. It is commonly recommended, with a sharp-pointed bistoury, to puncture the skin upon the inner or tibial side of the tendon opposite the internal malleolus, or higher, if the heel is very much elevated, and having withdrawn this to pass a probe-pointed bistoury between the tendon and the tibia, and while the tendon is made very tense by the hand of an assistant holding the foot, to cut the tendon by pressing the fingers upon it, thus crowding it upon the knife. If any shreds remain undivided, the fact is known by the failure of the heel to come down, and the bistoury is again partially withdrawn and passed under them, when they are divided by the same process by which the main portion of the tendon was cut. The reason for passing the knife on the tibial side of the tendon, is the less danger of wounding, by the point of the knife, the posterior tibial artery, which lies upon the inner side, and the same reason exists for cutting towards the skin instead of passing the knife between the tendon and the skin, and cutting toward the bone. A small piece of plaster laid over the minute incision, is all the dressing that is necessary.

It is common to describe instruments peculiarly constructed for this purpose, but they are unnecessary. Many of the instruments made for tenotomy are too delicate.

A common pocket bistoury is the only indispensable instrument, but there is a convenience in having a cutting edge only for about an inch from the point, in order not to have to pay attention to the skin at the point of entrance of the knife to avoid making too large an incision.

Apparatus for extension is immediately applied by some, but in order to secure union of the divided ends of the tendons, by organizing exudations, it may be most safe to postpone this for a few days, and then to make the extension very gradu-

ally. It is not known that the tendo Achillis, divided subcutaneously in early life, in the human subject, has ever failed
to unite; but in an experiment which I made, some years ago,
upon a dog, the divided tendo Achillis united only by shreds
of its investing sheath, which indeed may never have been
divided.

Many operators have been in the habit of dividing all the
tendons which afforded resistance to the reduction of the deformity. The tendon of the tibialis anticus has been, next to
the tendo Achillis, most often divided; being associated with
contraction of the muscles connected with Achilles' tendon
in T. equino-varus, occurring more frequently than all other
varieties together.

The tendons of the peronei muscles have been occasionally
divided for T. valgus, and as the plantar fascia is uniformly
shortened in the arching of the foot attending T. equinus
and T. equino-varus, it has very commonly been divided in
connection with the division of the tendo Achillis. The division of these tendons, and of this fascia, is not usually attended
with any trouble from bleeding, but a case is related by Tamplin, in his work "On Deformities," in which the internal
plantar artery was wounded in a boy sixteen years old, in
dividing the plantar fascia. After the employment of compression for several weeks, the resulting aneurism was finally
laid open, and the artery secured by a ligature. To avoid
this accident it is usual to pass the knife in under the fascia
on the inner side, so that the point of the knife may not by
any accidental dip, get beneath the artery.

The old rule of division of tendons was thus clearly expressed by Tamplin as applied to T. equino-valgus (p. 104),
implying the division not only of the tendo Achillis, but also
that of the tendons of the peronei muscles.

"If there is decided contraction, and the foot cannot be
completely adducted by the hand, you must have recourse to
the division of tendons."

An example of extreme reliance upon tenotomy is afforded

in the Lectures on Club Foot by James Syme, Esq., reported in the London Lancet (American edition) for June, 1855.

Tenotomy is chiefly relied upon, while mechanical appliances are almost entirely dispensed with.

A little lint is placed upon the wound, and a figure-of-eight bandage **worn for a few days, and that is all.**

Mr. Syme, in this lecture, deplores the disposition of many **surgeons, and especially** those devoted to Orthopedic Surgery, to rely chiefly upon mechanical appliances, **and to** regard the division of tendons as subsidiary. He especially objects to Dr. Little, because he cuts so little and relies so much upon mechanism, thus perpetuating the old methods of treatment, to the discredit of the methods by cutting.

The latest published opinions in favor of division of tendons are those of Dr. Bauer (p. 34 of the little book already referred **to), where he says,** "**The active forms** of valgus necessitate the division of **the contracted** peronei muscles, or of the whole group of the abductors, as the case may be. This is at least indispensable in inflammation of the tibio-tarsal articulation. . . . . . . **It is** difficult to steady **the articulation** with mechanical appliances in paralysis of the entire motor apparatus of the foot, but it is completely impossible to do so when the malposition of the latter is maintained by retraction of the peronei muscles. We at least have never succeeded by any of the devised mechanical auxiliaries. Meanwhile, the **deformity** increases, and gradually compromises the bones of **the tarsus.** Between the two evils we have to choose, and we **think** that division of the contracted tendons is the lesser."

It is suspected that the uniform success of **division of the** tendo Achillis, as introduced by Stromeyer, gave an unmerited **estimate** of the importance and utility of the division of tendons **and muscles** in general. A reaction in this estimate **has led many to discontinue the practice** of dividing tendons, except in rare cases **of** remarkable obstinacy, while others **seem still to believe in tenotomy with** undiminished zeal.

Among the former is Mr. Richard Barwell, of London, who

says, in the preface to his little book, " I studied these maladies from the orthopedic point of view ; and while tenotomy
was almost a novelty in England, **was so charmed** with the
easy change **of form, which, after such an operation, could be**
produced in most distortions, that I became an almost enthusiastic admirer of the **procedure. After,** however, following **up**
carefully **a** large number of these cases, I was pained to find
in how many of them the deformity more or less returned, **in
how many a** different, an opposite distortion supervened ;
while power over the limb was actually injured **or** destroyed
in so large a majority of instances, that its retention appeared
absolutely exceptional."

It is the division **of these tendons which, like the peronei,**
run in long ligamentous grooves along the tarsus, which is
most objected **to.** It is claimed that the function of these
muscles is often permanently suspended by division, either by
not uniting, or by adhering to their sheaths, so as **no longer**
to be able **to act upon the bones into which they are normally**
inserted.

Mr. William Adams, of London, has **been investigating this**
subject during the last **few** years, **and has** dissected **twelve**
feet, in which tenotomy had been performed. The results of
these investigations have been published under the title, " On
the Reparative Process in Human Tendons." Mr. Barwell
has reduced these **results to tabular form,** which is quoted on
the following page :

TABLE FROM "BARWELL ON CLUB-FOOT," ED. 1863, ANALYZED FROM "ADAMS ON THE REPARATIVE PROCESSES IN HUMAN TENDONS."

No of Cases	Tendons divided.	Results observed.	Time lived after operation.
*I.	Tendo Achillis, Tibialis anticus,	Non-union of tibialis anticus, . .	4 days.
*II.	Tendo Achillis, Tibialis anticus, Tibialis posticus, **Flexor** long. dig.,	**Non-union of** tibialis anticus. " " " flexor longus digitorum.	11 days.
III. (right. left.)	Tendo Achillis, Tibialis posticus,	Tibialis posticus adhered to the bone.	23 days.
III.	Tendo Achillis, Tibialis posticus, Tibialis anticus,	Tibialis posticus was supposed to be but was not divided.	30 days.
*IV.	Tibialis posticus, Flexor long. dig., ...	Union to all surrounding parts. Non-union held together by shreds of sheath to which other tendons also adhered.	18 days.
V.	Tendo Achillis, Tibialis anticus, Tibialis posticus, Flexor long. dig.,	Tibialis posticus and flexor longus digitorum adhered together and to the bone.	6 weeks.
VI.	Tendo Achillis, Tibialis anticus, Tibialis posticus, Flexor long. dig.,	Tibialis anticus and flexor longus digitorum adhered together and to the bone—ends of tibialis anticus hung together by shreds of sheath.	6 weeks.

In the next five cases, in Mr. ADAMS's work, the tendo Achillis only was divided.

| XII. | Tendo Achillis, Tibialis posticus, Flexor long. dig., | **Non-union of** tibialis anticus, **Neither** retraction nor extension of the flexor longus digitorum. | Several years. |

ANALYSIS OF THE PRECEDING TABLE.

Division of the Tendo Achillis, 12 *Cases.*

United, in 12 cases.

Division of the Tibialis Anticus, 4 *Cases.*

United, in **1 case.**   Not united, in **3 cases.**
Adherent to surrounding parts, equally destroying the function of the muscle, in 1 case.

Division of the Tibialis Posticus, 7 *Cases.*

Not divided, in **1 case.**
United, in **3 cases.**   Not united, in **3 cases.**

---

* **In Cases** I, II, and IV, **the time** was certainly too short to determine the ultimate result.

Adherent to bone or surrounding parts, suspending the function of the muscles, in 3 cases; that is, in all cases of non-union.

Division of the Flexor Longus Digitorum, 5 cases.

Union, in 1 case. Non-union, in 4 cases.

Adhesion to surrounding parts (among the cases classed non-union), in 2 cases.

From this analysis we may well hesitate before dividing any tendon about the foot, except the tendo Achillis. If the result in these cases is of any value, the division of these tendons should only be practised in instances in which, from permanent loss, or paralysis of the opposing muscles, a permanent loss of muscular contraction is desirable in the muscles whose tendons are to be divided.

The result of investigations of the processes of restoration after division of tendons, in cases in which the suppurative process is escaped, is well stated in the following quotation from Paget's Surgical Pathology (p. 179, American edition). The reparative material exuded around and between the ends of a divided tendon, is described as "presenting to the naked eye the appearance of a soft, moist, gray substance with a slightly ruddy tinge, accidentally more or less blotched with blood, extending from one end of the tendon to the other, having no well-marked boundary, and merging gradually into the surrounding parts. In its gradual progress, the reparative material becomes firmer, tougher, and grayer; the redness successively disappearing from the centre to the axis; it becomes also more defined from the surrounding parts, and in four or five days (in rabbits), forms a distinct cord-like bond of connection between the ends of the tendon, extending through all the spaces from which they have been retracted, and for a short distance ensheathing them both. The surrounding tissue loses its vascular appearance, the bloodvessels regain their normal size, the inflammatory effusion clears up, and the integuments become looser and slide more easily." . .

" With the increase of toughness, the new substance acquires a more filamentous appearance and structure. After the fourth day, the microscope detects nuclei in the previously homogeneous fibrine-like material, and after the seventh or eighth day, there appear well-marked filaments like those of the less perfect forms of fibrous tissue. Gradually perfecting itself, but at a rate of progress which becomes gradually less, the new tissue may become at last in all respects identical with that of the original tendon." . . . . . . "In the specimens presented by Mr. Tamplin, Achilles' tendon and the tendons of the anterior and posterior tibial muscles of a child nine months old, in whom, at five months old, all these tendons were divided for the cure of congenital varus, the child had perfect use of its feet after the operation, and when it died, no trace of the division of the tendons could be discerned even with the microscope." . . . . . "Like a scar, the tendon contracts and becomes firmer. It is, therefore, impossible to say what length of new material was, in these cases, formed into exact imitation of the old tendon. However little it may have been, such complete reparation is exceedingly rare."

The organization generally appears less perfect than in the cases above mentioned, presenting a distinct boundary to the touch.

The newly organized material has great strength. In the experiments detailed by Paget, the divided tendon of a rabbit six days after division, sustained a weight of ten pounds. A specimen experimented upon ten days after division, sustained a weight of fifty pounds, but gave way with a weight of fifty-six pounds.

The following interesting observations and experiments, by Dr. L. T. Hewins, of Loda, Iroquois County, Illinois, show the influence of young age upon the activity of cicatrix-formation, to connect the divided ends of tendons, or to pull them together.

Upon a dog four years old he failed. Upon dogs ten days old, and three months old, he succeeded after removing portions of tendons. He also succeeded perfectly on a rabbit.

He observed the reproduction of tendon, or substitute for it, in the extensor digitorum manus in one man thirty-five years old, three-fourths of an inch having sloughed off, and in another man aged thirty-eight half an inch having been lost by sloughing.*

These latter cases were successes under difficulties, the wounds being open and granulating, and presenting the conditions most favoring the agglutination of the tendons to the bones and other surrounding parts. The influence of motion in elongating adhesions, and reducing shapeless masses of newly organized material to the shape and function of tendon, whether permanent by the persistence of the new material, or temporary, by its gradual shortening and disappearance, is well illustrated.

A very remarkable case is reported† by Dr. Warren Webster, Assistant Surgeon, U. S. A., of rupture of the tendo Achillis; several months afterward securing union by cutting the ruptured ends square off, and retaining the ends in contact by two sutures of strong silk.

The tendo Achillis was ruptured while running a foot race. Three months afterward, there was "an intervening gap between the divided ends about one inch in length, where but little plastic matter seemed to have been poured out to fill up the space."

The operation consisted in exposing the tendon by a free dissection, removing the interposed substance, cutting the extremities of the tendon square off, and retaining them in apposition by two interrupted sutures of strong silk. Extreme flexion of the leg and extension of the foot were maintained for six weeks by a slipper upon the foot, a dog collar around the thigh, and a cord intervening. The leg was bandaged, to lessen the action of the gastrocnemii. A high-heeled shoe was afterward worn for some weeks. The fourteenth week after the operation, the patient walked with scarcely any lameness, and the tendo Achillis appeared to be perfectly united.

* Amer. Med. Times, Sept. 3, 1864, p. 117.
† Transactions Illinois State Medical Society, 1864, p. 84.

M. Bouvier is quoted by Barwell, as having divided, in a dog, in 1842, the flexor carpi radialis, the flexor carpi ulnaris, the flexor digitorum sublimis, and the flexor digitorum profundus. In none of these did the subcutaneous wound unite so as to restore the use of the parts. In another experiment, the tendons **did** not unite at all ; in another, the several structures were massed together. M. Bouley met with the last result in an experiment upon a horse.

It is probable, that in some of these cases of massing together, there would be afterward an absorption of portions of organized exudations, which impede the movements of tendons, like that which occurs after a general union of tissues in the neighborhood of fractures, so that the result finally, would **not be** quite as bad, as might be inferred from these statements.

An objection strongly urged, even to the division of the tendo Achillis, is that the " cicatrix-contraction" which attends all solutions of continuity, united by the interposition **of** extensive organized exudations, gradually diminishes the distance between the cut extremities of the divided tendon, so that they are finally brought nearly or quite together. This makes a bad compensation for the advantage gained at first, by the necessity for the wearing of apparatus to prevent the recurrence of the deformity, while this process of cicatrix-contraction is going on. In the treatment without tenotomy, the muscles are from the first made to grow longer, by a change of their nutrition induced by the force gradually and persistently applied, rendering the progress at first more slow, while in the treatment by tenotomy, this growing of muscles to a greater length has afterward to be secured, when the cure fallaciously seems to have been completed, and perhaps after the case has passed from under the supervision of the surgeon.

The following is Barwell's language on the subject :

" The reunion of the tendo Achillis, after its division for talipes equinus, is almost a certainty, but it" (the division) " permanently weakens the muscles, nor is such a procedure, as a rule, an efficient cure of the disease ; partly because the

gastrocnemius and soleus are not the principal muscles affected, and generally have very little to do with the malposture ;
partly, because contraction is sure to recur." (P. 120.)

Notwithstanding all this, however, there are occasional instances in which, even Mr. Barwell, anti-tenotomist as he is,
would divide the tendo Achillis.

"I do not mean to deny that, occasionally, when there is
either great want of development, or great degeneration, it
may be necessary to divide the tendo Achillis, but it should
always be avoided if possible, since it is merely a temporary
expedient, which always leaves behind it a certain deformity."
(P. 127.)

In contrast with this again is the language of Bauer (p. 24):
"**As a general** thing, you have only to deal with the contracted muscles, and *division* is the sovereign remedy. But if
the case has existed from infancy, the bones have in form
accommodated themselves to their abnormal position; the tibio-
tarsal articulation is crippled; then the prognosis is rendered
doubtful, and the case may be irremediable."

"It is a common observation of orthopedic surgeons, that
the relief of contracted muscles by tenotomy reacts most favor-
ably upon the nutrition of the affected extremity, and nutritive
supply promotes, self-evidently, its growth and development.
Passive motion coöperates in the same direction."

A question of interest here arises, as to what part the division of the tendo Achillis takes in the restoration of the muscular function.

It is probable that the movements of flexion and extension,
which attend **the** treatment following the division of the tendon, and subsequent to the reunion of the divided **tendon,**
graduallly induces a lengthening of the muscular fibrils, and
this lengthening is a necessary condition to their shortening,
under the irritation of electricity, or of any other irritant.

This opposing force should be either elastic or alternating,
in order to obtain the most stimulating effect upon the muscles
in process of restoration ; permitting the frequent exercise of
contraction, with yielding force, so graduated as to restore the

length of the muscles in their passive state, upon the decline
or cessation of their contraction.

The alternating movements of the tendons of the still paralyzed antagonist muscles, first pushing and then pulling these
tendons, and in a minor degree pushing and pulling the muscles themselves, invite a flow of blood to the muscular substance, favoring its continued healthy nutrition, and the earliest possible revival of nervous power, when the paralyzing
cause, residing in the brain, in the spinal cord, or in the course
of the nerves, whether from organic lesion or sympathetic
action, is removed.

If the cause of the paralysis is such a destruction of nervous
substance as to result in complete and permanent paralysis,
the alternating movements of the muscles will at least tend to
preserve their volume, by keeping up their nutrition, by making
it mechanically possible for the blood to circulate through all
their capillaries ; motion being as essential to the freest circulation through the muscles as through the lungs.

The general health has then the benefit of a well-distributed
circulation, in addition to the local advantages of attention to
this indication.

Yielding Force.—The plan of yielding force, called by Dr.
Henry G. Davis, " elastic extension," is very properly denominated by him the " American plan," and to him is due the
merit of having been the first to employ it systematically,
and with a full appreciation of its value; acting in a manner
similar to that of muscles, alternating in the extent of their
movements with the changing degrees of resistance to be
overcome.

Apparently from ignorance of American medical literature,
Barwell claims this plan as his own. This is one of the instances in which several claimants for originality may be
equally honest and original, the merit, however, consisting in
the application of some other invention, which makes a revolu-
tion of the given art, not only easy but unavoidable.

In this case, the invention at the bottom, is the manufacture
of elastic rubber, placing in every one's hand, a most facile

means of meeting an indication which the older surgeons saw, but had **no ready** means of accomplishing. (See Trans. Am. Med. Assoc., **1863**.)

The most obvious considerations in connection with the treatment **of** cases of talipes, a great majority of which **are** congenital, arise from the anatomy of the **parts**.

The short interosseous ligaments, the capsular ligaments, **and** the synovial membranes, are adapted to the abnormal relations of the bones to each other, and these cannot be reached by the knife nor broken by any mechanical appliance safe to apply. To overcome the distortion perpetuated by the condition of these interosseous textures, mechanical appliances, capable of producing a constant tension upon them, must be worn **until the** changed nutrition of the parts secures increase in the length of some, and diminution in that of others. This may require months, and even years, but is sure to be achieved if persevered in sufficiently long. The operation upon the long tendons, saves the time otherwise necessary to be consumed in the elongation of the muscles of which they form the **attachments**, allowing the tension to come at once upon the short tendons, muscles, ligaments, and membranes. The division of tendons may be chiefly relied upon in those cases not congenital, in which the sole cause of the deformity and its perpetuation is the contraction of the long tendons, but in all the cases, embracing nearly all the congenital cases and some of those not congenital, in which some of the ligaments of the tarsal joints are too short, all operations without persevering mechanical appliances will be found worse than useless, disappointing all concerned, and producing in the public mind distrust of the resources of surgery.

It is astonishing that a surgeon of the age, experience, and reputation of Syme, should fail to make any distinction between these two classes of cases, allowing himself to be not only deceived, but the cause of many a disappointment in inexperienced operators, who may be led by his advice to rely upon the knife in cases in which the cure is to be attained only by the diligent application of mechanical apparatus for weeks and months.

In cases of obstinate **resistance to** reduction by extension, the progress can be greatly facilitated by the occasional application of force, while the patient is insensible from the influence of ether.

The same condition is then artificially produced, which occurs in a subluxation or sprain. The most tense ligamentous fibres are torn without a complete rupture. The investments of the muscular fibres, in the shortened muscle, are either slightly torn interstitially, or put upon extreme tension. All this is followed by increased vascularity, which is **favorable** to change of tissue, in obedience to the tension afterward applied to it, for the purpose of elongation.

This has been a common practice among American surgeons for many years, though Barwell, strangely enough, claims it as his peculiar **invention.** He says, with much apparent satisfaction (p. 116), " This is also a procedure of *my own* adaptation to these diseases, and is one from which very great advantage may be drawn." He very properly goes on to say, " I would limit its employment to severe cases, and would caution surgeons against the use of violence ; since, when once the muscular power is annihilated by the anæsthetic, very little force is required to place the foot in a normal position."

Electricity.—Electricity has been employed to remove the condition of the muscles upon which the deformity has been supposed to **depend.**

The philospohy of its action is sufficiently explained by the notice, on page 39, of the power of electricity to preserve the **volume and** force of muscles through the contraction and movements secured by the excitation of this agent. The experiments of Dr. Reid, upon paralyzed legs of frogs, are a sufficient illustration.

It is obvious, that the interrupted current should be applied to the paralyzed or weakened muscles, for the purpose of inducing muscular movements, and the improved nutrition to be secured by the increased supply of blood, and, consequently, the more liberal presence of the materials of nutrition.

As previously explained, however, the interrupted current may be applied in powerful shocks, sufficient to paralyze muscular irritability, to the contracting muscles in those cases in which, according to Todd and Brown-Sequard, the rigid, cold, and bloodless muscles are indicative of central nervous irritation. The theory in these cases is, that the irritation which causes the involuntary and permanent contraction of the striped voluntary muscles, acts at the same time upon the unstriped involuntary muscles of the bloodvessels, inducing such a persistent contraction as to prevent the flow of blood in sufficient quantity through the contracted tubes.

In this case, the electric excitation is employed, not so much for the purpose of securing movement of the voluntary muscles, as to paralyze the muscular walls of the bloodvessels, and render nutrition possible by the more free supply of blood. To obviate the cerebral excitement resulting from this method of employing electricity, the patient might be rendered anæsthetic by the inhalation of ether or chloroform. Much of the discrepancy of opinion with regard to the employment of electricity may be cleared up by this distinction. What has in practice been found important, can hardly be denied in the face of the strong advocacy of those who have seen its good results. Thus Bauer, in enthusiastic language, says:

"The most efficacious remedy in behalf of innervation is electricity. It should be used with assiduity every day, and for months in continuation. It will stimulate the existing mobility, and prevent structural decay." . . . . . "Electricity is the substitute for volition, and the best local gymnastic agent." Dr. Bauer quotes some cases, in which there was apparently paralysis of all the muscles of the leg, in which the energetic and persevering application of electricity proved of great benefit.

It is obvious that when, by unresisted tonic contraction, the muscular fibres and their fasciæ have shortened to their utmost, neither electricity nor the prick of a pin can make them shorten any more. In this state of extreme and permanent contraction, the muscle has degenerated from its function of producing motion, to that of holding parts in position, which is the function of ligaments and the condition of *contracture*. A galvanic current can make no impression which is known by movements, because this agent and other irritants only produce contraction. If, however, the muscular fibrils and their investments are first made to grow longer, **by frequently repeated** pulls upon them, or by constant force varying in intensity, thus restoring the muscle to a greater **or less** extent, to the possibility of performing its natural function; then, after so much progress has been made towards the cure, it might be expected that electricity would index it by the contractions which would result from its application.

It is difficult to see, however, on what rational principle electricity should be applied to the shortened muscles (which at the same time that they are shortened, have good nutrition indicated by temperature and volume), with any other intention than to determine whether they could shorten any more, or to ascertain, in the progress of treatment, in a case in which a muscle had been shortened, and degenerated beyond the possibility of exciting contractions by the electric current, whether any progress had been made, or, perhaps, to throw light upon **the probable** replacement of the muscular substance by fatty degeneration. In the latter case, electricity could not produce movement.

With the acknowledged power of electricity in securing exercise and nutrition, in muscles whose circulation **is repressed** by rigidity or retarded by the nutritive inactivity of paralysis, such is the difficulty of selecting the case for its employment, as to induce its entire rejection by some practitioners of eminence.

Dr. R. B. Todd says (Lectures on Paralysis and Diseases of the Nervous System, American edition, p. 152):

" You will often be consulted as to some expedient for promoting the restoration of paralyzed limbs to their normal condition. To this question, after having given a fair trial to the various means which **have been proposed for** this purpose, I must reply, that I know of nothing which more de-cidedly benefits paralyzed limbs than a regular system of exercise; active when the patient is capable of it, passive if otherwise.

"As to the use of electricity, which is now much in vogue, or strychnia, which has been recommended, I feel satisfied, as the result of a large experience, that the former requires to be used with much caution, and that the latter is apt to do mischief, and never does good. I have seen cases in which, after the employment of electricity for some time, that agent has apparently brought on pain in the head, and has excited something like an inflammatory process in **the brain**. And so strychnia **will** also induce an analogous condition of the **brain, and** *will increase the rigidity* of the paralyzed muscles. Some good may occasionally be effected by the use of friction or cold water, or shampooing, all of which tend to improve the general nutrition of the nerves and muscles."

Mechanical Treatment.—The successful treatment of talipes and other deformities upon philosophical principles, is usually considered one of the triumphs of modern improvement.

The following quotation, however, shows that the Father of Medicine perfectly comprehended the true principles, and made the best possible use of the materials which the civiliza-tion of his time afforded.

Hippocrates (on Club-foot, in his Treatise on Articulations, writing five **hundred** years before Christ, quoted from Biggs' " Orthopraxy"), says:

"Wherefore, then, some of those displacements, if to a small extent, can be reduced to their natural condition, and especially those at the ankle-joint. Most cases of club-foot are remedial, unless the deviation is very great, or unless the affection occurs at an advanced period of youth. The best plan then is to treat such cases at as early a period **as** possible, before the deficiency of the bones of the foot is very great, and before there is any great wasting of the flesh of the leg. There is more than one variety of club-foot, the **most of** them being not complete dislocations, but impair-ments connected with the maintaining of the limb habitually in a certain position.

" In conducting the treatment, attention must be paid to the following **points: To push** back **and rectify** the bone of the leg at the ankle from without inward, and to make counter pressure on the bone of the heel in an outward direction, so **as to bring it into line, in** order that the displaced bones may meet at the middle and side of the foot, and the mass of the toes with the great toe are to be inclined outward, and retained so; and the parts are to be secured with cerate, containing a full proportion of resin, with compresses and soft bandages **in** sufficient quantity, but not applied too tightly ; and the turns of the bandage should be in the same direction as the rectifying of the foot with the hands, so that the foot may appear to incline a little outward ; and a sole made of leather, not very hard, is to be bound on, and it is not to be bound to the **skin; but when you** are about to make the last turns of the bandages, and when it is all bandaged, you must attach the end of one of the bandages that are used, to the bandages applied to the inferior part of the foot on the line of the little toe, and then this bandage is to be rolled upward, in what is considered to be a sufficient degree, above the calf of the leg, so that it may remain firm when so arranged ; in a word, as **of** moulding **a** wax model. You must bring into their natural position the parts which were abnormally displaced and contracted together, so rectifying them with your hands and with **the bandaging in** like manner, as to bring them into position, not by *force*, but *gently*, and the bandages are to be stitched, **so as** to suit the position in which the limb is placed; for different modes of deformity require different positions ; and a small shoe made of leather is to be bound on external to the bandage, having the same shape as the Chian slipper.

" But there is no necessity for it, if the parts be properly adjusted by the hands, properly secured with bandages, and **properly** disposed of afterward. This then is the mode of cure, and *it neither requires cutting, burning, nor any complex means ;* for such cases yield sooner to treatment than one would believe. However, they are to be fairly mastered only by time, and not until the body has grown up into natural shape. When recourse is had to shoes, the most suitable are the buskins, which derive their name from being used for travelling through the mud; for this sort of shoe does not yield to the foot, but the foot yields to it. A shoe shaped like the Cretan is also suitable."

Of this remarkable description, Dr. Adams, the translator, remarked: "Now it appears to me a lamentable reflection, as proving that valuable knowledge, after having been discovered, may be lost again to the world for many ages; that not only did subsequent authorities down to a recent period, not add anything to the stock of valuable information which Hippocrates had given on the subject, but the important knowledge which he had revealed came to be disregarded and lost sight of, so that, until within the last few years, talipes was regarded as one of the *opprobria medicinæ.*"

Modern Plans.—The plans of treatment adopted by the pioneers in modern orthopedics are still retained by many of our surgeons of reputation. Some immovable and inelastic frame of wood or iron, properly padded, was employed to bring the foot around into proper position; the apparatus being changed for another of different shape as the restoration progressed, or adapted with joints to change with the changing shape of the foot.

The simplest and oldest form is a flat splint, to apply to the leg, with a flat, thin wood-piece, the edge of which was fastened upon the end of the splint, in the form of a cross, upon which the foot and leg were bound by roller-bandages. In contrast with the simplicity of this, are the complicated machines invented by Scarpa, Scontetten, and others, in the beginning of the great awakening upon the subject of orthopedics, about thirty years ago.

Scarpa's shoe has an iron sole, an iron heel-piece at right angles with this, and a brace running up the leg, while a spring, attached to the side of the shoe, gives a pull with some elasticity for straightening the incurved foot; all this is properly padded, and provided with straps and buckles. The

vertical brace passes up on the projecting or convex side,—
upon the outer side in talipes varus. The illustration, Fig.
74, shows the iron framework of this complicated machine.

FIG. 74.   FIG. 75.

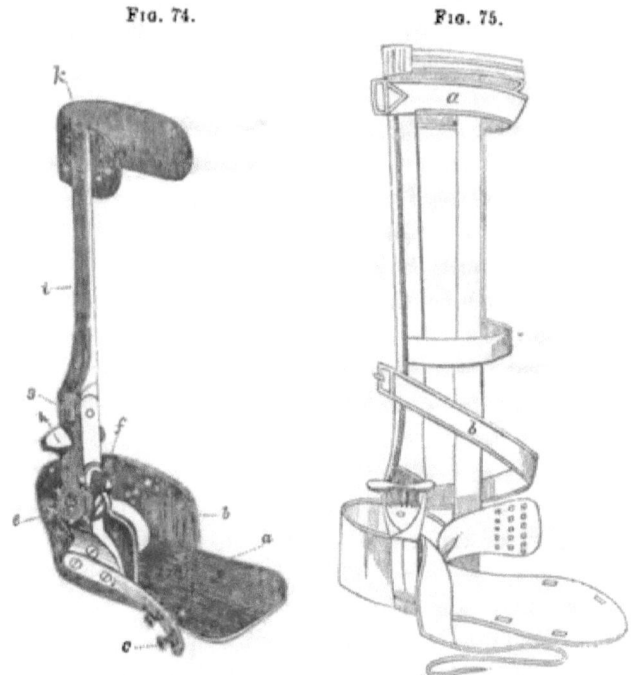

Explanation.—The shoe is in a straight position. *a.* The sole. *b.* The semicircular portion to
embrace the heel; a portion behind is cut away. leaving a hole for the end of the heel to pro-
trude. *c.* The horizontal spring for the abduction of the foot. *e.* A hinge in the upright
portion. *f.* A triangular screw-head which is turned with a key, and causes the point of the
instrument to turn down. *g.* Another hinge. *h.* Another triangular screw-head, which,
being turned with a key, bends the foot part outward. *i.* The upright shaft or brace. *k.* The
semicircular part to go round the leg, and act as a fixed point of the apparatus.

Scontetten's apparatus differs from Scarpa's chiefly in
having two shafts, one passing up on each side of the leg.
Fig. 75 illustrates it without all its padding.

Dr. Bauer, in his work already so often quoted, employs a
slight modification of Scontetten's apparatus as the utmost
advance in the art at the present time.

These machines, however, are not well adapted to any
species but T. equinus and T. varus, and for each varying

size of foot, an expensive apparatus must be made. They are uncomfortable, extremely liable to produce ulceration, almost destitute of elasticity, acting chiefly upon the ankle-joint, and moving the foot as a whole, failing to move the tarsal joints upon each other, as is done when the foot is grasped by the hand. They are difficult to make, except by skilled instrument makers.

The use of the starch bandage, the gypsum bandage, and gutta percha, for retaining fractured bones in position, suggested the employment of these means in the treatment of talipes. In the use of the starch and gypsum, the bandages are applied in the same manner as for fracture, and the foot, thus enveloped, must be held in position by the hand of the operator, until the wet and yielding shell becomes dry and stiff.

In the employment of gutta percha, as recommended by Dr. A. C. Post, of New York (Fig. 76), the material is cut to re-

FIG. 76.

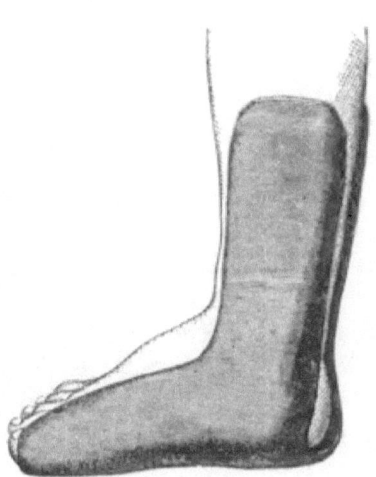

Method by gutta percha, according to Dr. A. C. Post. Medical Record, vol. I, No. 1.

semble somewhat the shape of half the upper leather of a boot, and there may be a single side, or two sides, cut from a single sheet of gutta percha, one-fourth to one-third of an inch thick, or the sides may be from two pieces. These are placed between **two layers of** muslin, dipped in hot water and kept there until perfectly yielding (and no longer, for the **stuff will melt), then applied,** invested with a roller bandage, after which the foot is held by the hand of the surgeon or his **assistant, until** the material is cool enough to yield only as it **is elastic.** Or **we** may **go back to** Hippocrates, and employ leather, which is to be applied wet, carefully bandaged, and the foot held in position until the leather becomes dry and stiff.

The use of adhesive plaster, introduced **for** this purpose about the year 1850, was a great advance in the art. The method consists in cutting strips of convenient width and long enough to envelop the foot and pass up the leg nearly to the knee, there to be fastened in place by circular strips passing **round the leg,** over which the upright strip (or strips, for there **must usually be several of** them), are turned **so** as to clinch **them to prevent their** sliding.

For **T. varus** the plaster ascends **on the** outside, and **for T.** plantaris and **T. valgus on the inside, and** for **simple T. equi-** nus on both sides. **It is sometimes found convenient to carry** the fastening above **the knee for greater space for application of the** plaster.

This expedient holds the **foot in the position** in **which it is placed by the** hand of the surgeon, except a little **sliding that plaster** will always be **guilty of. The desideratum is a method which is** within **the skill of every person of** ordinary **ingenuity, to be** made **of materials** always at hand and free **from** expensiveness, and which **will** not only hold the **parts in the position** in which they are **placed by the hand, but gain something;** as would be accomplished by the hand, **if it were constantly** employed.

It occurred to me, as it occurred independently to many others, soon after **the introduction of** elastic rubber for varied use, that a piece **of elastic rubber ribbon could** be interposed in the vertical strip of adhesive plaster, so as not simply to hold the foot in the position in which it was left by the hand, but to be constantly gaining by a yielding but unintermitting

stretch night and day, gradually elongating the opposing muscles and ligaments, and by the slight mobility attending the
elastic rubber, permitting some passive motion in the muscles
assisted by the elastic appliance, whereby their circulation is
increased, with a more rapid nutrition and a more speedy accommodation to their altered length of contraction and relaxation.

The method of interposing rubber ribbon between two pieces
of adhesive plaster, is very well shown in this figure (Fig. 77)
from Andrews, illustrating the treatment of T. varus. It is
convenient to make the rubber adjustable by buckles, which
are not shown in the cut.

This was for some time supposed to be the last advance of
which the art was capable, but, ul-
ceration sometimes occurred upon
the edge of the foot, where the circulation was too much impeded by
the circular compression of the
plaster around the foot. There
seemed to be a lack of some expedient by which the fold of the
tarsus could be straightened out,
so as to restore the foot to its normal breadth. An obstinate case,
attended with ulceration of a delicate skin, led me to devise an
appliance which is a tolerable substitute for the hand; but before
describing it, a few pages must be
devoted to the plan of treatment

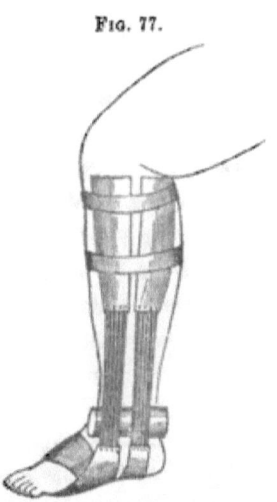

Fig. 77.

For Talipes varus (from Andrews).

pursued by Mr. Barwell, to explain which, his book (on Club-
Foot, &c.) seems to have been chiefly written.

The peculiarity of Barwell's plan consists in his method of
attaching the proximal end of his tension apparatus, which is
this: Starting with the idea of making the artificial tension,
the exact complement of that of the partially paralyzed muscles, he aims to act as nearly as possible upon the same bones

to which these muscles are attached (and in the same direction), by adhesive plaster fastenings, while the points from which the pull comes, are the origins of these muscles.

Thus, for T. varus, the fastening is made on the exterior anterior side of the upper part of the leg, at a point over the origins of the peronei muscles, in such a way as to get two-thirds of the length of the leg for the position of the rubber spring upon which he relies for the pull.

The lower attachment is made to imitate as nearly as possible the insertions of these muscles; but for retention to the skin, the lower adhesive plaster passing downward over the cuboid and fifth metatarsal bones must cross the bottom of the foot, and fasten upon the inner side above the sole. In order to get a retention of the rubber spring upon the upper part of the leg, a broad strip of adhesive plaster, twice the length of the leg, is applied over the course of the peronei muscles, over the fibula, and upon this, a piece of tin, a little narrower than the plaster, is laid, and the lower end of the plaster turned up over it, so that the inside (or sticky side) is outside, for adhering to the roller that applies round the whole, to hold it fast. The upper end of the tin is turned over from the leg, and has a hole punched in it, and into this hole an eyelet is inserted; a similar eyelet is inserted in the adhesive plaster which passes across the bottom of the foot, and between these is stretched a rubber spring. By the combination of two or more of these expedients, he is enabled to obtain tension which imitates the combined action of the peroneus longus and p. brevis, passing behind the external malleolus, and the peroneus tertius passing in front.

For talipes valgus, he makes a similar appliance on the inner side of the leg and foot, to supply the deficiency of the partially paralyzed tibialis anticus and tibialis posticus. The pull must here be in two directions, as in the other case.

In talipes plantaris (flat-foot), he makes a direct lift upon the hollow of the foot, by an anterior appliance compensating the deficient lift of the tibialis anticus.

In talipes equino-dorsalis, he makes also a direct lift further

forward. He explains this deformity as being the direct opposite of talipes plantaris or flat-floot, in which the medio-tarsal
joint sinks too low, hence it must be lifted up; while in talipes
equino-dorsalis, the same joint rises too high, while by the contraction of the tibialis posticus, the peroneus longus, the p.
brevis, and the flexor longus digitorum, the metatarsus is
flexed or drawn down, bringing the toes to the ground, while
again the instep or "waist," of the foot rises too high. **He**
thinks the action of the sural muscles, through the tendo
Achillis, upon the calcaneum, a minor element in the deformity,
and hence a particular objection to the division of **the tendo**
Achillis, in addition to **the general objection** arising **from per-**
manent injury to the tendon.

The account would be more nearly correct to say, that in
addition to the contraction of the tibialis posticus and flexor
longus digitorum, the foot is arched too high by the shortened
condition of the abductor pollicis, the flexor brevis digitorum
perforans, the abductor minimi digiti, and the musculus accessorius, with shortening of the plantar fascia, to correspond
with this disproportionate contraction of these muscles.

The pull directly in the line of these tendons, besides being
a refinement of treatment, difficult and sometimes impossible to
execute, is one which acts at a great mechanical disadvantage,
implying a greater pressure upon the skin, to accomplish a
given amount of change of position, **than** would be required
by a direct pull.

If it had been the design of nature to **make only slow move-**
ments of the extremities, **there would** have been nothing
gained by binding **down the** tendons under transverse ligamentous substances as they pass the joints. A much **smaller**
force would have accomplished the purpose, by acting in a
straight line between the origin and the insertion of **any**
muscle. The facility of movement and grace of form secured,
by giving the tendons oblique attachments, are elements unnecessary to be regarded by the orthopedist. There is this
great disadvantage **in** this attempt to imitate the oblique
action of the muscles: that the pressure upon the skin is three

or four times what it is necessary to make it, when the most
direct pull is obtained. The importance of gaining the most
power with the least pressure upon the skin of the foot can
hardly be exaggerated. Ulceration of the foot, where the
pressure applies, is the greatest difficulty which it has been
the study of surgeons to avoid.

It cannot be said that the muscle which is partially paralyzed is more assisted by the oblique pull than by the direct,
for the passive motion of the muscle is communicated by the
push and pull of the tendon; and this to and fro movement
must be the same for a given amount of motion of the parts
to which the tendon is attached, whether the movement is
effected by an oblique pull in the direction of the attached
end of the tendon, or by a power acting at a less mechanical
disadvantage, like the hand of the operator, or any apparatus
which acts in a similar manner.

Illustrations of Barwell's Method.—Fig. 78 shows the manner of applying the plaster over the tibia, and the tin over it,
and the plaster under the sole of the foot for T. plantaris : *b,*
a trapezoid piece of plaster, into which an eyelet has been
fixed, adhering to the sole of the foot, to act as the insertion
of the tibialis anticus tendon ; *d,* a strip of plaster adhering
over the tibialis anticus muscle, and having its lower end
hanging down more than the length of the limb. The letter
d is upon the upper portion of this free part. *c.* A piece of
tin carrying at the top a wire loop, *f.* The free end of the
plaster is turned up on the tin, and a roller applied to hold
all fast.

Fig. 79 shows the process completed. The lower end of
the long piece of plaster has been turned up over the lower
end of the tin, and in the figure, circular investments of plaster
are shown instead of a roller. *g.* Strip of plaster surrounding the foot, but leaving out the end of the plaster *b,* having
an eyelet in it. (See *b,* in Fig. 78.) *l.* A rubber spring running from this eyelet in the plaster, which comes from under
the sole of the foot, up the leg, to the wire loop, at the upper
end of the tin.

Fig. 80 shows the application of the same plan in the treat-

ment of T. varus. Two springs are shown, imitating the
action of the peroneus tertius in front of the external malle-

Fig. 78. Fig. 79.

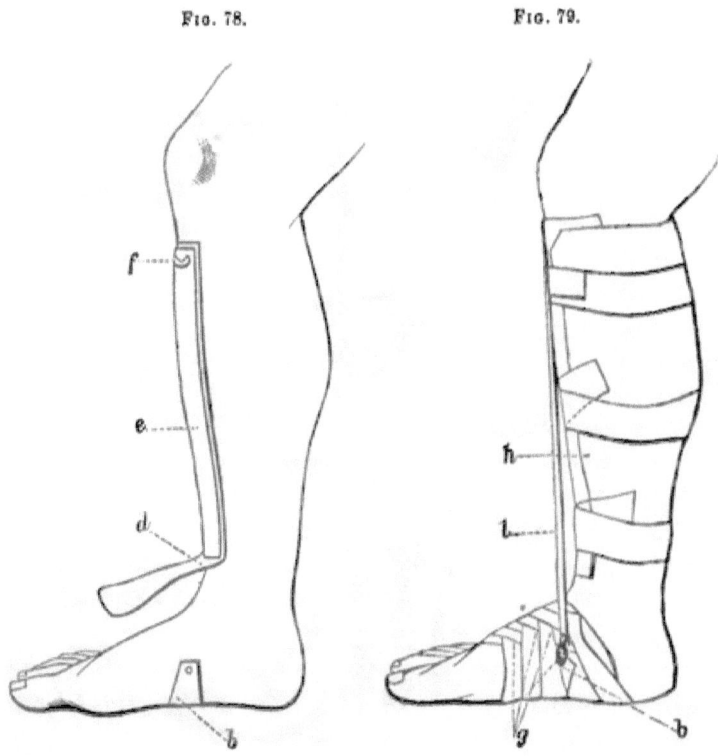

olus, and of the peroneus longus, and the **p. brevis behind**
the malleolus.

m. A trapezoid piece of plaster applied across **the** bottom
of the foot, and having an eyelet, like *b*, in Figs. 78 and 79.
The course of this, under the circular strips, is marked by dotted
lines, *n.* It is represented as being split, so as to embrace the
fifth metatarsal bone. *n.* The eyelet, for the attachment of the
rubber spring by a piece of catgut, or other strong cord. *o.*
Circular strapping, covering but one piece of tin, placed just behind the fibula, with its layer of plaster on either side. *v.* The
remainder of the longitudinal strip of plaster brought down and

adherent to the circular ones. *t.* A rubber spring **assisting** the peroneus tertius. *u.* A rubber spring assisting the p.

FIG. 80.

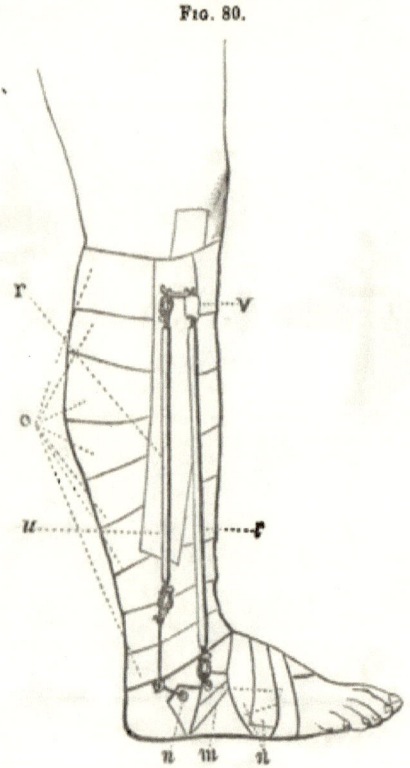

long. and p. brev.   **At the lower** part of the attachment of the spring, **marked** *u*, **is an** arrangement for changing the direction of the force, by an attachment around the limb. *v.* A short piece of rubber tube covering a hook, by which the spring is attached to the eyelet in the upper end of the tin. All the attachments are covered in the same way in practice to shield the hooks from the clothes.

In obtaining a pull from a space directly over the elongated muscles, by the plaster and tin appliances, a very considerable pressure is produced over the whole circumference of the part. We know that a moderate pressure, like that produced in health by the skin and fascia, and by a laced stocking, when these are relaxed in varicose veins of the extremities, is favorable to muscular tone, but a greater degree of **pressure**, like that produced by ligating a member for cramp, is **unfavorable** to muscular contraction.

The requisite degree of adhesion can be secured without binding the limb so tightly as to interfere with the free circulation in the muscles, and their necessary expansion in contracting.

It is probable that this method of attaching adhesive plaster will come into general use **wherever it is desirable to gain space** for elastic extension.

The application of **adhesive plaster to the** foot, as employed by Barwell, **does not** materially differ from the method **for** many years **in** common **use**. The plaster cannot be stuck to the skin as the tendon is stuck to the bone. It **must** have a considerable breadth of attachment, or it will slide off. **This** necessary extent of surface cannot easily be obtained upon the foot without carrying the plaster round upon the opposite edge, so that its pull must approximate the bones of the metatarsus and of the phalanges. This force is the direct opposite to that which is produced upon an inverted club-foot (talipes varus), by walking upon it. The weight of the body, in walking, comes upon the cuboid, the fifth metatarsal bone, and corresponding phalangeal bone, **until, by** folding and twisting, the foot **is still** further turned, so **as** to bring **the** weight of the body **upon its dorsum.**

The plaster takes hold of the opposite **or inner border (in** talipes varus), and passing under the foot and **up on the out-**side pulls in the opposite direction. In all this there is no tendency to take the longitudinal fold or doubling out of the foot. The force simply untwists the malposition of the cuboid in relation to the calcaneum, and the cuneiform bones in relation to the scaphoid, and more than all the others, the scaphoid in relation to the astragalus. To the extent of the tilting of the astragalus in the ankle-joint, and the sliding of the calcaneum upon the astragalus, these deviations are also corrected.

It is obvious, by a glance at the skeleton, that an important agency in reducing the slight dislocation of the cuneiform bones upon the scaphoid, and the principal dislocation of the scaphoid upon the astragalus, is the unfolding of the foot to give it transverse breadth. This is one of the most important indications in cases in which the patients have been some time walking. It is easy enough to answer this indication with the thumb and fingers, taking hold of the foot and twisting it in directions opposite to those of the distortion; but the thumb and fingers soon tire out. It is possible to employ a succession of hands for that purpose, and this would probably be as effectual as any more artificial method. The desideratum is the invention of apparatus which will do what the thumb and fingers can do, and do it without tiring out, and without danger of producing ulceration from the persistency of unyielding pressure. The device to answer this end, without much expense, and in a method so easy of execution that it can be readjusted every day or two, is simply thus:*

For a patient ten years old, take a sheet of gutta percha one-third of an inch thick, or a sufficient number of thinner sheets to make that thickness, long enough to encircle the foot, and wide enough to extend from the middle joint of the phalanges to the medio tarsal articulation, i. e., the joint between the scaphoid and astragalus above, and the cuboid and calcaneum below.

Apply upon both surfaces of the gutta percha, an investment of muslin of good strength, and lay the whole, thus prepared, into a pan of water nearly boiling hot. While the softening process is going on, the foot should be wrapped with a roller, protecting the prominent points with pledgets of lint or cotton.

As soon as the gutta percha is thoroughly softened, it is taken out, still lying between its muslin investments, and so applied that its ends come together on the outside of the foot (in talipes varus), where the two extremes of gutta percha

* Gutta percha, in sheets, suitable for surgical purposes, can be procured of Bishop, 201 Broadway, New York.

should be welded by pressure between the thumb and fingers, previously dipped into cold **water** to keep the material from sticking to the fingers.

In talipes valgus, the extremes of gutta percha meet and project on the inner or median side of the foot. While the material is yet warm and yielding, a square piece of pasteboard is laid upon the dorsal surface of the foot, with a corresponding piece of oiled-silk or rubber-cloth underlying it to prevent its softening by the moisture of the wet muslin investment, and a similar piece of pasteboard is applied directly opposite upon the plantar surface.

A common pair of calipers, with screw fastening, is then applied, so that one leg rests upon the pasteboard upon the dorsal, **and the other** upon the pasteboard upon the plantar surface. The screw is then turned to secure very firm squeezing between the opposing points. This compression is continued until the gutta percha has become hard and unyielding, except by its elasticity. After this, the calipers are removed.

A hole is then punched through the projecting gutta percha, alongside of the metatarsal bone of the little toe in varus, and of the great toe in valgus. Into this hole a cord is inserted, which is fastened to a rubber ribbon or piece of rubber tube or cylinder, which must again have its attachment above by adhesive bands below the knee, above the knee, or by a padded roll to the pelvis which is thereby encircled. This last is the least troublesome attachment as it can, at any time be slipped off and put on again. In the last method, a knee-cap is necessary to make the tension-cord follow the angle of the limb in walking and sitting. The appliance to the foot should be removed and reapplied every day in hot weather, **and every alternate** day in cold weather, to avoid excoriation from pressure and retained exhalations.

The pressure if too long applied to a part, without intermission, favors absorption with ulceration; or, if acting with sufficient force, the death of the compressed parts, resulting in sloughing; while the moisture from the skin, with the ammonia which it contains, favors a softening or solution of the cuticle, thus increasing the natural sensitiveness of the parts to pressure.

In employing this appliance for T. varus, in which there is little of that elevation of the heel to make it a T. varo-equinus, the first part of the treatment consists in straightening the foot, elongating the plantar fascia and associated muscles, tendons, and ligaments. After this is well accomplished, the force comes upon the tendo Achillis, to elongate the great triceps extensor pedis. For this purpose, it is convenient to attach two springs to the gutta percha shell, one of which pulls upward as shown in the figure, while the other is carried backward and upward spirally around the leg, to **be attached** on the inside of the leg near the knee to the same fastening which holds the other spring.

I have lately adopted the plan in T. varo-equinus, of paying very little attention to the depression of the heel, or what is the same thing, the elevation of the metatarsus and phalanges, thinking that the process goes on better to attend first to the elongation of the plantar fascia and associated short muscles **of** the foot, thus securing an eversion **of** the front half of **the** foot upon the back half, the hinge **being** in the medio-tarsal articulation in front of the calcaneum and astragalus, thus converting the case into a talipes equinus; after which the elevation of the foot and the depression of the calcaneum is easy enough. This is done by attaching a rubber ribbon of pretty strong pull to the outer side of the gutta percha shell, and carrying it spirally around the leg, behind; then **on** the inner side, next in front, and then partly to the outer side **just below the knee, where** it is fastened by adhesive plaster attached above the knee.

As much power as may be desired is thus acquired, according exactly with the traction of the peroneus longus and p. brevis. The foot is by this means straightened, untwisted, and converted into a T. equinus, and when the heel is subsequently brought down by the elongation **of the** triceps extensor pedis, the symmetry of the **foot is** perfect, instead of presenting a high instep. The disposition of the foot to return to its former shape is much less than is the case when the front half of the foot is first elevated, with the plantar fascia still too short.

Figure 81 illustrates the method of applying the apparatus in talipes varus, to secure tension upon the pelvis.

1. Rubber spring. 2. Buckle for adjustment. 3. Gutta percha investment of the foot, to the outer side of which the tension apparatus is attached. 4. Projection of the toes beyond the investment, and above the application of the lower leg of the calipers, applied upon a piece of pasteboard, to secure sufficient distribution of pressure. 5. Calipers showing the screw by which the squeezing of the middle portion of

FIG. 81.   FIG. 82.

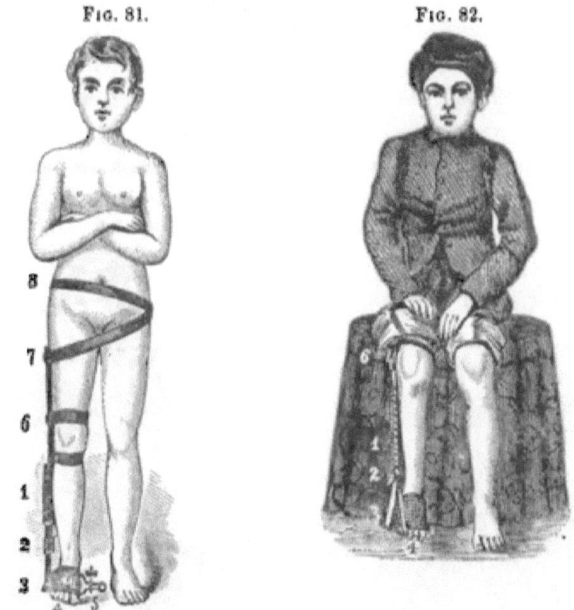

the gutta percha is produced. 6. Knee-bands. 7. Band to which the tension cord is attached, passing obliquely across to the opposite ilium. 8. Band around the pelvis to hold the other band from slipping down.

Figure 82 illustrates the same method with an attachment above the knee. It is convenient to have a secondary fastening below the knee, which is not shown in the cut.

The figures refer to the same parts as in the preceding cut. The calipers are supposed to have been removed, and the apparatus to have been fully adjusted. The whole may be inclosed in a moccasin or slipper, to enable the patient to walk about. If the patient is an infant, a stocking may be drawn over the apparatus. This method of attaching a rubber spring to the foot by means of a gutta percha investment, can very conveniently be combined with Barwell's method of adhesive plaster attachment to the leg.

Figures 83 and 84 are accurate copies of photographs of a case of talipes varus, in a boy nine years old, before treatment, and at the conclusion of treatment, at the end of three months. The flattening down of the tarsus is more perfect than can often be secured without the vertical compression of the foot in the manner just explained. The foot appears shorter than that of the other side, because in the deformed state it had fallen behind the other in growth, but the treatment has spread

FIG. 83.

FIG. 84.

Before treatment.   After treatment.

the foot out effectually, so that there is no danger of a recurrence of the deformity, without a nervous derangement capable of producing it from the first.

The following quotation from Barwell, p. 183, on Club-Foot, &c., aptly illustrates the effect often produced by a theory in hampering one's natural versatility, and driving him to awkward and difficult **expedients. The** quotation is in explanation of the difficulty of getting room upon an infant's leg for application of plasters, in a child aged six months:

"A little more difficulty" (than usual) "had arisen from the greater adduction of the foot; this rendered it difficult to fasten on so small a thing as a child's leg and foot, the plaster representing the peroneus brevis, so that the end to which the catgut was fixed did not come against the eyelet in the tin representing the pulley. This is **a** difficulty which always occurs in children's cases. I find it best overcome by cutting the plaster which is to represent the tendon of a Y shape, stretching it in the hand that it may not give way on the limb, turning down one of the ends, leaving it very short, and fastening the **eyelet into it**, while the other two ends are made to adhere, one on the sole and one on the dorsum of the foot, leaving **the inner** metatarsal bone uncovered. In these cases, **also, in spite of** any difficulty in knotting it, the catgut must be tied very short; the spring too must be as short as possible."

In this, Barwell recognized, without mentioning or explaining **it,** the evil of that folding influence upon the foot in talipes varus, arising from pressure of the plaster upon the first metatarsal bone. To avoid this, he stops his dorsal and plantar plasters short of meeting on the tibial side of the foot.

His practical difficulties are very much increased by his theory of getting his pull from over the partially paralyzed muscles. In talipes varus involving an elongation or loss of action of the peronei muscles, he must get his traction from over the fibula; and he is confined to the length of that bone, because these muscles have only their origins within this space. By carrying the attachment above the knee there is found plenty of room for the rubber spring, allowing something for the inevitable sliding **of** the plaster.

By adapting the gutta percha appliance to the foot, a firm fixture is secured, equal to a hand continuously applied, which not only does not increase the abnormal transverse doubling of the foot, but helps to flatten it out, thereby much facilitating the rotation of the top or tibial margin of the foot inward or downward, and the bottom or fibular margin outward or upward.

The origin of this theory was in a correct appreciation of the philosophy of the subject, **and the** failure of the most complete success grew out of too close an imitation of nature, where power is lost to gain rapidity of movement and beauty of form. In the artificial removal of deformities, rapidity is only the desire of inexperience, and beauty is out of the question; while it is of the utmost importance to avoid all **unnecessary** pressure upon the skin to which the appliances are attached. The more direct the pull, in imitation of the hand of the operator, the lighter will be the pressure upon the skin, the less the discomfort to the patient, and the more practicable the employment of as much force as the muscles and ligaments will bear without pain in these parts.

The fundamental idea which is at **the foundation** of my plan **of treating** talipes, is the invention and application of apparatus in imitation of the action of the human hand.

Iron shoes, and all cumbrous inelastic and expensive machinery, **are thrown** away. The restoration of the proper form of the foot is more likely to be the conclusion of the treatment when the muscles, tendons, and ligaments have been elongated without division, by the slower process of growth from **nutrition,** than when they have been factitiously elongated **by** division of tendons, and the interposition of cicatricial **material,** which will gradually contract to its complete disappearance. The plan here explained makes it practicable **to** avoid division of the tendo Achillis, in cases in which it might **be** necessary by the old methods, and even by the improved plans of Barwell.

After the treatment is completed, it is useful **to** steady the foot by a brace running up the side of the **leg,** having a joint exactly opposite the centre of motion in the ankle. The lower **part is** made of soft iron, so that the shape can be easily altered, and it is riveted to the sole of a common shoe by two copper rivets, the heads being placed **inside the shoe.**

The part above the joint, is a flat spring, conveniently cut from a worn-out saw. The yielding of this spring permits lateral motion at the ankle-joint, while the joint in the apparatus permits flexion and extension. At the top of the spring brace, which should reach about four-fifths of the distance from the ankle to the knee, a cross-piece is fastened, made of thick tin or thin iron, of the length of half the circumference of the leg, which serves, when bent to the shape of the leg, to prevent the brace from sliding backward and forward. Over the whole length of the elastic portion of the brace above the ankle, a leather investment of the circumference of the leg and brace is adapted, which is supplied with eyelets to lace upon the opposite side. The brace is always placed upon the side from which the deviation proceeds. The pull is, therefore, from the brace, so that there can never be any chafing of the skin against it. This saves all necessity for cushioning it. The brace is always supporting the ankle-joint, and always yielding as the foot treads upon uneven ground. The figures will make this description more intelligible.

In figure 85 all portions of the metal above the ankle are invested by the leather, but in the cut they are represented as being on the outside.

Fig. 85.

This apparatus will do very well for weak ankles, but should never be trusted, after treatment for talipes varus, as long as the instep is in the least too high. The foot should first not only have the twist entirely taken out of it, but if a T. varus it should not be left in the least degree a talipes dorsalis. It is entirely practicable, by the method here described, to convert it into a T. plantaris; but this is neither necessary nor desirable. After this thorough removal of the deformity, the surgeon is not likely to be afterward troubled with the case on account of a tendency to a return of the deviation, unless there should be a return of derangement of innervation, such as originally produced it.

It may be noted in closing, that in young infants, previous to walking, and before the infolding of the transverse diameter of the foot from the weight of the body upon its outer margin, the use of the gutta percha clamp is not very important. The adhesive plaster investment is usually sufficient, but the use of the elastic rubber ribbon is indispensable to satisfactory progress. Where the single ribbon is too delicate, its strength can be increased by doubling. It is convenient to attach a buckle or hook at each end of the rubber ribbon and to work the adhesive strips into them from above and below. The facility for adjustment is then complete.

In order to obviate the lateral pressure of the plaster upon the foot, a sole of leather may be first applied under the foot, made a little wider than the sole of the foot, and the strips of plaster wrapped around this and the foot combined, as is practised by Dr. H. G. Davis, of New York.

It seems to me that any case of talipes, in a patient under fifteen years of age, ought to be restored ; but a continuance or a repetition of the derangement of innervation, which originally produced the deformity, may tend to reproduce it, requiring the continued use of an elastic aid to the enfeebled muscles, which may be worn inside of a boot, not differing in principle from the appliances already described, though more delicate and less bulky.

It is not supposed that perfection has yet been attained in this art, nor is it wise to be satisfied with the improvements already made, nor to believe that there is as much known about it now as there ever will be. If, however, we could see what improvements are to come next, we should immediately make them. Experience feels out the future, but sees the past with eyes open.

Deviations at the Knee-Joint.—To avoid the necessity of minute explanations of treatment, which can be readily inferred from what has been said upon the subject of talipes, it is convenient to introduce here, *Genu Valgum* (knock-knees), which is usually a congenital deformity of not very rare occurrence, depending upon defect of the lateral ligaments of the knee-joints, requiring the division of the external lateral ligament or its elongation by mechanical force perseveringly applied, and mechanical aid for a long time after the restoration seems to be complete. If the treatment is undertaken before the period of walking, it is not important that the apparatus should have a joint opposite the knee, and the splint can be extemporized with great ease.

After the patient begins to walk, however, there should be a joint opposite the knee, and if the apparatus is connected with the shoe, there should, also, be a joint opposite the ankle.

This figure (Fig. 86) from Bigg, will afford a very complete conception of the proper apparatus.

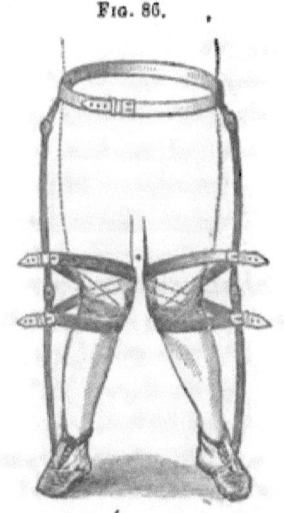

Fig. 86.

Apparatus for knock-knees (from Bigg).

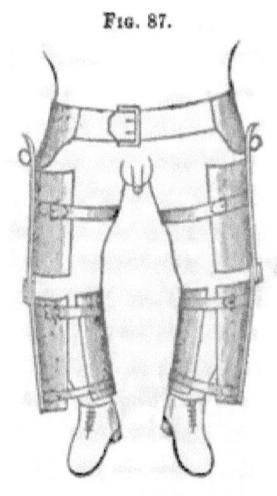

Fig. 87.

Apparatus for bow-legs (from Andrews).

Bow-legs: a deviation in the opposite direction, whether depending upon the elongation of the external lateral ligament with contraction of the internal lateral ligament and associated tendons, or upon curvature of the bones from premature walking, or softening of the bones, requires a similar method

14

of mechanical treatment, the splint being placed upon the inside, or as seen in the cut (Fig. 87, from Andrews), upon the outside, exerting pressure instead of traction upon the knee.

UPPER EXTREMITY.

The upper extremity is the seat of malformations and malpositions, though less frequently than the lower, and the necessity for restoration is equally great. The general principles and plans of treatment are the same.

Elastic force constantly drawing, yet yielding to the muscles when they act, permitting thus, such movements as keep up the flow of blood, and such changes at the seat of pressure, as to relieve all parts from the danger of ulceration and sloughing.

There is one peculiarity of treatment applicable to the upper extremity, that is, the necessity for making distinct traction upon individual thumbs and fingers, while it is hardly ever necessary to make application to individual toes. It is always easy to get an attachment upon the arm and fore-arm, but it is not easy to attach the apparatus to the fingers in extreme distortion.

The employment of finger-stalls, which may be fingers of gloves, affords attachment to the elastic material which is relied upon to change the position. It will often be found exceedingly convenient to make incasements for the fingers of gutta percha, to which the rubber spring may be attached. For the greatest success and to avoid ulceration, the gutta percha stall should be removed every day, and after warming sufficiently to change its shape, it should be reapplied. In this way the local pressure is not only relieved, but the pressure is, with each dressing, somewhat varied. Where obstinate resistance is to be overcome, it is of the greatest importance to keep the parts in a condition to bear the full amount of pressure. This, however, must never very far exceed the ability of the patient to bear the pull without pain.

The importance of this precaution is all the greater, **on ac-count of the** necessity for relying upon traction without the aid of tenotomy. Besides the danger of wounding vessels and **nerves** in subcutaneous sections of tendons, there is the addi-**tional danger of** their not uniting when once they have been **divided.**

The length of the grooves in which the tendons move, gives them the same liability to non-union, which attends the division of the long tendons connected with the foot.

CICATRICES.

It is proper to notice, in this connection, those deformities most frequently occurring in the face and upper extremities, **resulting from loss of** substance by burns, abscesses, and areolar erysipelas, in which the forearm becomes permanently flexed upon the arm, the wrist flexed or extended, so as to be turned backward or forward along side of the forearm, or one or more thumbs and fingers flexed into the palm of the hand or doubled back upon the phalanges.

If the loss of substance is not too great, very much may be done in the way of prevention, by splints and adhesive plaster, aided by elastic rubber interposed to compensate for the sliding of the plaster, as explained in the treatment of talipes. The attempt should be made to keep the member in the straight position, in order to force the sound tissues to yield in the di-rection of the circumference instead of gliding longitudinally as they are most inclined to do.

PLASTIC OPERATIONS.

In some situations, a permanent line of immobility **can be** interposed between the contracting cicatrix and the movable parts, which cannot be protected from the gradual but irre-**sistible traction by any other means.**

This expedient may be practised in case of a burn upon the neck, or upper part of the chest, resulting in depression and eversion of the lower lip. An incision may be made along the lower border of the lower jaw-bone, and the periosteum dissected up, so that the integument of the chin becomes afterwards continuous with the cicatricial material in contact with the bone. The traction of the cicatrix below thus comes to expend itself upon the lower jaw. The powerful masseter and temporal muscles are usually capable of counteracting the tendency to depression of this bone, and the integument of the chest gradually yields to the contraction of the cicatrix.

From the repulsion of parents and guardians to the resort to a measure which itself constitutes a deformity, this expedient will usually be postponed until a considerable degree of deformity has already been produced.

It is fortunate that the line of immobility can be as readily established in an operation for the removal of deformity, as in one for its prevention. The disappointment arising from the tendency of the cicatrix, incident to a plastic operation, to contract and reproduce the deformity attempted to be removed, can sometimes be avoided by this precaution. Nothing short of peeling up the periosteum should be relied upon for this purpose.

Where important muscles, vessels or nerves would require to be divided, in order to secure this immobility, the measure is of course impracticable. In these cases, and where there is great destruction of tissue, the malposition of parts will gradually progress, in spite of the utmost attention, and nothing will save the member from permanent distortion, but the implantation of new tissue from some other part. From the nature of the case, after great loss, the new supply must come from the opposite arm, the trunk, the thigh, or from another person, and the danger of sloughing of the flaps in this class of plastic operations, is so great that one must be prepared for disappointments, delays, and failures.

DEFORMITIES AFTER FRACTURES.

THESE result from loss of a portion of the fractured bone, which may occur at the time of the injury as in gunshot fractures; from subsequent exfoliation or necrosis; **from want** of coaptation of the fractured surfaces during the period in which reunion is taking place; from subsequent contraction of the new-formed **bone, as** occurs in the cicatricial material of soft parts; from bending at the place of fracture, in consequence of muscular contraction, the weight of the member, or of force applied while the new bone is in a yielding state; or from **delay or failure of** the fragments to unite, allowing them to deviate **from** their proper directions, after the retaining apparatus ceases to be worn.

It seems, that in a few exceptional instances, a resoftening of the bony callus occurs, permitting a bending in obedience to muscular contraction and other forces; thus, Dr. F. H. Hamilton has reported a case, occurring under his own observation, of a boy eight years old in whom a very decided outward bending of the femur occurred **six** months after the fracture and four months from the time the patient began to walk. The progress of this bending spontaneously ceased.*

Prevention.—The prevention of lateral deviation following fractures, implies the employment of mechanical means, not only during the period of six or eight weeks while consolidation is being effected, but for a considerable period afterward, in cases in which, from unequal loss of substance **upon different** sides of the bone, or from faulty coaptation, an unequal amount of new bone has been produced, resulting, necessarily, in subsequent unequal cicatrix-contraction.

This prevention may in some cases be altogether impossible,

---

* Transactions American Medical Association, 1857, p. 272.

but a recognition of the principle, and of the liability to disappointment in the final result, may be an aid to the surgeon in guarding his own reputation.

The prevention of lateral deviation incident to delay or failure of union, implies the *securing of union*. This makes it appropriate to discuss the methods **to** this end, in this chapter, which can be most conveniently done in connection with the account of the means of removing the deformity.

Without attempting to discuss the treatment of fractures, it may be mentioned, that deformity by shortening, is of course unavoidable after loss of any considerable portion of the shaft of a bone, and that in many cases in which, after oblique **fractures** without very close coaptation of the fractured surfaces, the length of the limb is preserved by extension during treatment, subsequent shortening must occur from the tendency of the new-formed bone to contract.

This is the same process by which cicatrices in soft parts contract and approximate adjoining parts—and by which the redundant portions of new bone (or callus) disappear.

The progressive shortening of limbs after the occurrence of union and the cessation of treatment, in oblique fractures of the femur, can only be explained upon the supposition of progressive contraction of the newly formed bony material, just referred to in connection with lateral deviation.

Correction.—The correction of lateral deviation connected with bony consolidation may be attempted:

1. By force gradually applied through a considerable period, for the purpose of securing a change of nutrition in the substance of the new bone, and a consequent change of direction. This is the principle on which bow-legs are straightened in the growing period, and the deviations incident to rickets are attempted to be corrected. This method requires great care to avoid excoriations of the surfaces, and great perseverance, **for** the change must be very gradual.

2. By more rapid bending during an early period, while the bone is susceptible **of** *interstitial* fracture, without complete solution of continuity. This corresponds with the bending of

the long bones of children. A change of form is produced
without any angular projection, and which is more or less permanent, unless the original shape is restored by force.

3. By refracture at a later period, partial if the consolidation is only just too great for bending, and complete, if a
greater degree of brittleness has been attained. After this,
the case is the same as that of an original fracture.

4. By section or resection, or section combined with fracture. The conditions of compound fracture are necessarily
attendant upon this method, involving suppuration and long
confinement.

5. Softening of the bone by perforation, and consequent inflammation ; in order to make it possible to change the direction of the bone by bending, by interstitial fracture, or by
partial fracture. It is by softening the bone in connection
with the inflammation excited, that the seton introduced by
Dr. Physick, is employed as an aid in correcting deformity.

Nelaton (Elemens de Pathologie Chirurgicale, tome i, p.
682) says, that " Reflecting that a seton passed through a
callus, is capable of exciting inflammation and of softening it;
Weinhold perforated it by means of a trephine of very small
dimensions, and secured to the member the length which it had
lost. Notwithstanding this success, I do not recommend the
method, which is exceedingly severe, and perhaps no more
certain than the preceding."

Dieffenbach, in 1845, introduced his method of drilling and
driving ivory pegs into the orifices and leaving them there.
This was of course followed by many of the conditions of
compound fracture, with occasional exfoliation.

In 1854, Dr. Daniel Brainard, of Chicago, published (Transactions American Medical Association, 1854), a detailed account of experiments upon animals, and cases in men, in which
he had secured softening of bone by drilling *subcutaneously*.
The idea was borrowed from Dieffenbach's plan. The attempt
to avoid suppuration, **and yet** secure all the benefit which
accrues from the suppurative form of inflammation, is the distinctive improvement ascribed to Dr. Brainard.

The method consists in drawing the skin somewhat to one side, and introducing a drill from one-eighth to one-fourth of an inch in diameter, and perforating the bone a greater or less depth. This is done in several directions. About ten days are then allowed to elapse before attempts are made to change the direction of the bone. E. R. Bickersteth, Esq., Surgeon to the Liverpool Royal Infirmary, has returned to Dieffenbach's plan, only leaving the drill in its place in the bone, instead of withdrawing it and introducing ivory pegs.*

The introduction of this means of softening bones and rendering them capable of changing their direction without extreme violence, and with greater certainty with regard to the place where the yielding will take place, greatly lengthens the period proper for interference. Thus, Norris, in a note to Fergusson's Practical Surgery, page 341, states as an extreme limit that the curve has been reduced as late as the one hundred and twentieth day from the injury. Oesterlen did not attempt refracture later than the twenty-fourth week, and Bosch made six and a half months the latest period at which it should be attempted with a screw. Skey refractured the tibia at the end of thirteen months, in a boy aged fifteen.†

DELAY OR FAILURE OF UNION, WITH OR WITHOUT DEFORMITY.

While fractures sometimes unite under the most adverse circumstances, at other times union is delayed or does not take place where the appearances are at first most favorable. This difference of results, independent of external circumstances, can only be accounted for by the assumption of constitutional differences of aptitude to bony formation. While ossification will sometimes extend through an inch or two of plasma, reaching from one fragment to another, the separation to the extent of one-fourth of an inch will, at other times, prevent the bony

---

* See Lancet, March 19, 1864; Braithwaite, No. 49 (1864), p. 77; Chicago Medical Examiner, July, 1864 p. 416.

† T. Holmes, in Holmes' Surgery, vol. i, p. 808.

union of the fragments. While privation and starvation some-times fail to retard union, there is in some constitutions a ne-cessity for a liberal diet, to afford the necessary stimulus to bony deposit. The antiphlogistic remedies for high inflammation, if continued unnecessarily long, may sometimes prevent union, while in other instances no practical amount of local or gen-eral reduction will interfere with bony formation. While, there-fore, it is never safe to omit any of the conditions of success in the treatment of fractures, the greatest number of unfavor-able circumstances may be insufficient to cause failure if the ossific tendency is strong.

It is suspected, that a fuller investigation will show **that** separation of the fragments to a distance of one-fourth of an inch or more from each other, and insufficiently nutritious diet **at the period** of from three to five weeks from the date of injury, are the most frequent causes of delay or absence of union. If this shall be affirmed by experience, it will follow that the two most important points for the surgeon to attend to are, the ap-position of the broken surfaces of the fragments, and the pro-per nourishment of the patient during the ordinary period of ossification. It must not be forgotten, however, that extreme fulness in diet may beget conditions of the system more dan-gerous and unwelcome than protracted non-union.

The delay having occurred, and the fragments remaining beyond the usual period, connected by soft callus of a greater or less degree of firmness, the treatment will at once suggest itself: to secure local stimulation by frictions upon the skin; movement of the broken surfaces upon each other; a resort to more liberal diet; securing a better general health by exercise or exposure in the open air; and pressure upon the parts with reference to the approximation of the fragments when **this is** practicable, and when the delay may be suspected to depend upon the motion of the fragments upon each other, the diminu-tion or arrest of this motion.

All these failing, some means of inducing more active capil-lary circulation with congestion or inflammation must be re-**sorted to.**

1. In the list of means to this end, is the passing of a seton through the callus between the fragments. This may be supposed to excite inflammation in all the parts immediately surrounding the seton, including the neighboring periosteum. As an important point of treatment is to get the action of ossification started somewhere, in order to favor the propagation of this action through the fibrinous material constituting the callus, the treatment is based upon intelligible physiological principles.

From the known tendency of long-continued inflammation in and near the periosteum to induce bony deposit, it may be that Dr. Physick was right in retaining the seton a long time, with the result of a protracted congestion in the neighboring bone and periosteum. It may be in practice better to try the seton first for the short period, and if that fails, to try it for the long period where this method of treatment is pursued.

2. The injection of some stimulating agent like iodine into or around the callus is founded on correct principles, but must be so extremely uncertain, that in the possession of surer means, it is not worth any further trials.

3. Galvanism passed through acupuncture needles introduced into the substance between the fragments, or in close proximity to them, can only be expected to succeed by exciting hyperæmia or inflammation.

4. Opening the parts, and scraping or sawing off the ends of the fragments, converting the case into one resembling compound fracture. In very old cases, in which the false joint resembles a capsular ligament with its inclosed synovial membrane and cavity, this severe proceeding becomes necessary. In any case in which the duration of the false joint is not measured by years, it is not easy to conceive this process to be necessary.

5. Dieffenbach's method of drilling and introducing ivory plugs, leaving them there to excite suppuration, can hardly be conceived better than the seton carried through the soft callus between the bones, while the risk of necrosis must be a strong

objection to the proceeding. The plug may, however, be employed in cases in which it is extremely undesirable to carry the seton through the whole diameter of the limb, on account of the amount of suppuration and the degree of irritation necessarily following.

6.* The modification of Dieffenbach's method, followed by Detmold and Brainard, which consists in practising the perforation, by drawing the integument to one side, in order that it may slide over the orifice as the drill is withdrawn, thus increasing the chances of avoiding suppuration and the possible consequence, caries or necrosis, has two theoretic recommendations. First, a very great disturbance of the particular portions of the bone drilled is effected, giving rise to the production of new plastic material for the formation of callus in the track of the drill, without the occurrence of suppurative inflammation. Suppuration here, as in the healing of other tissues, must be supposed to retard the union, though the active capillary circulation in the vicinity of its seat may result in subsequent ossification. The case may be thus stated : If the bony deposit can be induced by congestion or non-suppurative inflammation, it is more speedy than that brought about by suppurative inflammation. Yet there may be cases in which a long-continued inflammation with suppuration will induce the formation of bone after the failure of a shorter course of inflammation without suppuration.

In the cases in which there can be success by congestion or non-suppurative inflammation, suppuration is an evil retarding the result. In the other cases it is a necessary attendant upon the prolonged inflammation.

Second. When the operation results in the effusion of plastic lymph without suppuration, there are new centres of ossifica-

* See New York Medical Gazette, &c., edited by D. Meredith Reese, M.D , Oct. 12, 1850, p. 232, and Transactions American Medical Association, 1854, p. 555. Detmold employed Dieffenbach's gimlet shaped-conical drill, but Brainard, the better to penetrate dense bone, employed the truncated shape, usually adopted in drilling iron and ivory.

tion in the chips of bone cut off by the drill. These are left in the track of the drill; some of **them in the soft callus** between the ends of the fragments.

That these minute fragments of bone become parts of the living tissue which organizes around them, is certain; for, if they did not, they would, by the offensive emanations of dead bone, excite suppuration and work their way to the exterior. The importance of these little fragments cut off by the drill, as centres of ossification, may have received too little **attention.** As in crystallization the introduction of a single minute crystal may be sufficient to start a process which is backward to commence without catalytic aid, so the process of ossification, when slow to begin, may be set in operation by a fragment of bone, or periosteum, imbedded in the plastic material. To obtain this advantage of the bony fragments it is, of course, necessary that suppuration in the track of the drill should be **avoided.**

7. Applying metallic wires around the fragments to approximate them and prevent lateral motion, answers an obvious **indication.** To apply a wire around the fragments, it is, however, necessary to convert a simple fracture into the condition **of a** compound fracture; and afterwards, when union has taken place, the wire is to be left in, or removed at the expense of much disturbance of the parts. If a silver, gold, or platinum **wire** becomes covered with organized lymph or granulations, **it** can do no harm, and may be allowed permanently to remain.

8. Perhaps a bone might be drilled through both fragments and held in apposition by a rivet of one of these metals. The **presence of the** rivet after the completion of the healing process would do no harm, and if a permanent discharge should be the result the metal could be readily removed.

9. Metallic points arranged for pressure on one or more of **the fragments for the purpose of** approximating them.

This expedient, where **the** nature of the parts makes it practicable, supplies an important indication. It accomplishes all **that can** be secured by the application of wires with more certainty, without extensively disturbing the soft parts, and the apparatus **is easily** tightened or loosened, increasing or diminishing the pressure, and is easily removed altogether.

Whether the separation of the fragments has been occasioned by muscular action, or by the interposition of muscle or other material, the pressure will be constant, tending continually to approximate them.

Malgaigne's spike for oblique fractures of the lower portion of the tibia, is intended to prevent what it may afterwards be employed to remove, *i. e.*, a too wide separation of the fragments. In this apparatus the counter **pressure is by means** of a strap passing around the limb, including a splint, which distributes **the pressure** upon the opposite **side.** In other cases, **the** counter pressure must be by means of opposing points, **acting upon the** opposed fragments, in order **to** bring them into close contact. Skill in constructing and adjusting the apparatus, will be chiefly exercised in making it occupy a sufficiently small space not to be in the way of placing the limb alternately in various positions, while the process **of union is** going on.

Wherever the application of pressure by metallic points, penetrating the soft parts and pressing the bony fragments together, becomes necessary, it would have been important to apply them in the first place, in order to bring the fragments into close contact, and to favor union by what is termed by Paget *immediate union*, or by *primary adhesion.*

It is found by experience, that very little **pain is occasioned** by wearing for weeks a steel point, applied with considerable force, to the fragment to be held.

The treatment does not convert a simple fracture into the condition of a compound fracture, for the point can be applied at a sufficient distance from the place of fracture to avoid this complication. When, however, points have to be applied to opposite sides of the limb to act upon different fragments at the same time, they must be nearly or quite opposite each other; but as it is only in oblique fractures that this treatment is admissible, it will in very rare cases be necessary to penetrate the interior lesion of the soft parts.

In cases of compound fracture, the points can be introduced into the wound or through the uninjured soft parts, as may be most convenient. This, as a first treatment of fracture, may be found to be less painful than apparently more comfortable **modes of dressing, obviating the movement** of one fragment upon the other by the closeness with which the surfaces are brought together. Some periosteal inflammation must be excited, which, if it extends **to** the fractured lines, can only the **more** certainly result in bony formation, whether as a primary treatment or as a method of curing non-union. A slight exfoliation of bone may occur at the spot where the metallic point is made to press; but this is a trifling consideration, in comparison with an increased efficacy in the treatment.

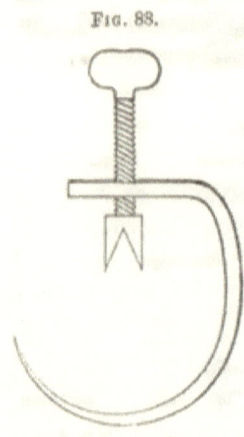

FIG. 88.

Modification of Malgaigne's spike, employed for delayed union in oblique fractures. The apparatus is arranged with two points, but in most instances a single point will do as well.

A single point **may** be applied by means of the metallic **yoke and strap,** as employed by Malgaigne, and where two or more points are to be applied on opposite sides of the limb, an apparatus may be constructed resembling the clamp used by ladies to fasten to a table any fabric for greater convenience in sewing upon it, or like some forms of tourniquet made **to apply** opposing pads by means of a steel yoke approximated **by** some screw arrangement. The pads would for this purpose be replaced by points. The apparatus should be so arranged as to be capable of compressing the fragments as closely as may be necessary to keep them in apposition, and to hold them without any yielding whatever. There should be **no elasticity in** the retaining apparatus. (**Fig.** 88.)

If the pressure of the fragments upon **each** other is found **to be painful to the patient, the screw may** be loosened a very little, as a very small relaxation of pressure will be capable of affording relief.*

CASE I.† *Non-union of Tibia with Deformity, unsuccessfully treated by Drilling; afterwards successfully treated by Drilling followed by Compression of the Fragments by means of Malgaigne's Spike.*—Lieut. Samuel L. Hamilton, Co. **F, Nine**teenth Regiment Illinois Volunteers, on the 15th of May, 1862, had both fibula and tibia of the right leg broken, a short distance above the ankle, **by** being thrown from a wagon, lighting upon his feet. IIe was treated in the **army hospital, and** the patient says his surgeons had considerable difficulty in keeping the bones in proper position.

After a few weeks, a starch bandage was **applied,** and the patient went upon crutches. The fibula united by bony material, but the tibia remained ununited. Some deformity existed from the action of the muscles, sliding the lower fragment upon the upper and bending the fibula, bringing the outside of the foot to the ground.

Operation under Ether by Drilling, after Brainard's Method, Nov. 5, 1863, five months and twenty days from **the date of** the injury: The fragments of the tibia were forcibly moved upon each other, and two holes were drilled through both fragments and the intermediate soft callus. The callus seemed, from the jumping of the drill, to be a quarter of an inch in thickness.

A side splint was applied, extending from the upper portion of the tibia over the malleolus, to which the limb was firmly bandaged. The fibula thus received the whole force of

---

* My friend, Dr. William S. Edgar, of Jacksonville, recently surgeon to the Thirty-second Regiment Illinois Infantry, having, during the past winter, a case of troublesome oblique fracture of the lower **third** of the tibia, applied first, a metallic **spike, and** afterwards substituted for it an ivory spike, and liked the latter much **better.** Whether the irritation upon the surface of the bone is greater or less, he thinks **the ivory** more agreeable to the fancy of the patient.

† These cases and illustrations were published in the American Journal of **the** Medical Sciences, Oct. 1863, p. 313, and in the Transactions of the Illinois **State** Medical Society, **for 1863, p. 72.**

the bandage on one side, while upon the other side, the force of the bandage was received upon the malleolus, and the upper portion of the tibia by the intermedium of the splint. In two weeks the constant pressure had straightened the fibula so that there was no deformity. There was no perceptible motion between the fragments, and the splint was directed to be worn some time longer with the expectation of success.

This operation proved a failure, and the movement of the fragments upon each other became obvious enough.

Second operation.—*Drilling and the Application of Malgaigne's Spike*, March 11, 1863, ten months from the injury, and four months from the previous operation : A very obvious deformity had been reproduced. The muscles acting upon the fibula as a fulcrum, had bent it, so as to bring the outer side of the foot to the ground, while the inner side was slightly lifted from it. The patient having been brought under the influence of ether, the fibula was forcibly straightened by interstitial breaking, or by bending with breaking of portions of the substance ; after which a quarter-inch drill was introduced between the fragments, passing from below upward and backward, and freely rotated in the space between the two fragments, breaking up the soft intervening callus. The fragments were thus shown to be one-quarter of an inch asunder. A small probe was introduced and left as the drill was withdrawn. Three holes were then drilled through the anterior fragment and intermediate callus and into the posterior or lower fragment.

The limb was then put upon a posterior splint, which was a double inclined plane, and the steel-point of Malgaigne's spike placed about an inch above the lower end of the upper fragment, through an incision made in the skin by a bistoury, the strap adjusted beneath the splint and the screw turned down until the probe left between the fragments was very firmly grasped by the approximation of the fragments. A light sidesplint was applied on each side within the yoke holding the spike. The probe was then pulled out from between the fragments.

With slight adjustments from time to time this apparatus was worn without removal twenty-eight days. The patient took opium enough during the first few days to quiet pain. He was overtaken with a chill, to which he had for several months been subject, after which he had the consequent fever with a pulse of 120. He took quinia for this, and lager beer. As soon as he was free from his ague, he discontinued medicine. Considerable swelling and suppuration occurred around the spike, which was not attended with much pain. The apparatus looked worse than it felt.

April 8th. The twenty-eighth day: Removed the dressing, and applied a tin side-splint.

17th. Applied a starch bandage, which was split on the 19th, and directed to be worn two weeks longer.

FIG. 89.

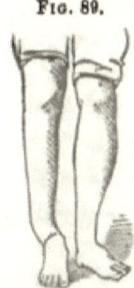

FIG. 90.

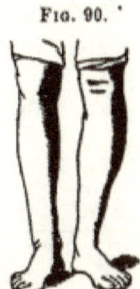

Appearance of leg of Lieut. Hamilton,
March 10, 1863.

Appearance of leg of Lieut. Hamilton,
June 27, 1863.

(Engraved from Photographs.)

There is a node on the inner side of the tibia, exactly opposite the point occupied by the spike, as if periosteal inflammation had extended around the limb from the point of irritation by the spike. Minute exfoliations afterwards came out in the vicinity of the point pressed upon by the spike. Consolidation followed this treatment, gradually imparting confidence to the patient, who cautiously ventured to walk upon the limb. The patient left to rejoin the army the first of July, and remained in service during his term of enlistment.

CASE II. *Ununited Fracture of Tibia and Fibula of three years' duration, with much Angular Deformity from Contraction of Muscles. Reduction of Deformity by Extension and Lateral Pressure—Drilling the Bones according to Brainard's Method, resulting in Bony Union without deformity or Lameness.*—Augustus Simpkins, of Pike County, Illinois, aged about thirty-five years, had a simple transverse fracture of the middle portion of the tibia and fibula of the right leg, by the fall of a tree.

There is said to have been much swelling and inflammation, and the skin was cut to let out the effused fluids. Cold applications were kept upon the leg, and the patient restricted to a low diet. No union by bone followed, and the angular deformity (the foot being carried out, making the leg look like a limb with a knock-knee) resulted gradually from muscular contraction. When the patient stands erect the toes only come to the ground, the lower portion of the leg being at an angle of 45° with the other leg.

June 12, 1861. The non-union has been of three years' duration. Applied the most powerful extension practicable by the lever arrangement of Jarvis's adjuster attached to the distal end of a long splint, the counter-extension being upon the ischium and groin, while lateral pressure was applied by a sort of tourniquet working with a strong screw.

Forcible working of the ends of the bones upon each other was practised by taking hold of the limb with the hands, and the tendo Achillis was divided. With all this, the limb was not restored to its straight position, and the apparatus breaking under the great strain applied, the process was stopped. The limb was dressed so as to retain as far as possible what had been gained.

FIG 91.

a. Screw with its concave pad applied to the projecting angle of the leg. *b.* Hook for retaining the screw, making counter-pressure upon the splints. *c.* Long splint, which is the medium of extension. *d.* Back splint attached to the long splint, for aiding in securing the counter-lateral pressure.

After five days, not much inflammatory excitement had appeared, and the limb was subjected to another process. The bones were drilled from one fragment into the other in six places, taking different directions, all traversing the soft callus between the ends of the bones. The extension and lateral pressure were applied as in the first instance, only with stronger apparatus. The extension was from the ankle, by means of a roller applied around it to hold the loops. The limb was not only straightened by this operation, but the muscular resistance was so completely overcome, that I bent the limb in the opposite direction without difficulty. The thigh, leg, and foot were then placed in a side-splint, made of tin, and kept in it until the consolidation was complete, except when taken out for washing and friction to the skin.

In three weeks from the first operation he went home, a distance of forty miles, riding about half the way in a buggy. The splint was worn about ten weeks. Perhaps it might have been laid aside sooner, but the patient after three years' experience, was afraid to trust his limb too soon.

During the operation, a mixture of ether and chloroform was inhaled, and to quiet the subsequent pain, morphia was freely administered. No other antiphlogistic treatment was resorted to than cathartics.

The result in this case should lead us never to despair of success, until after trials of all means of cure. As the fracture of the tibia was transverse, the interposed substance was subjected to great pressure by the contraction of the muscles, and there was no want of apposition to account for the non-union. It is suspected that the antiphlogistic treatment was too long continued. The fragments of the fibula became overlapped as the limb assumed the angular position, but when brought into proper relations by straightening the limb, the fragments became united by bony substance.

The following figures (Figs. 92 and 93) represent the conditions before and after treatment.

I am led to think, from my own experience, and from the published accounts of cases in which the callus alone has been punctured, that the perforation of the callus by awls or drills, which do not penetrate the bony substance, is useless, and perhaps worse than useless, by breaking up its organization without influencing the bone and periosteum, whence the process of bone-formation most readily proceeds.

Fig. 92.   Fig. 93.

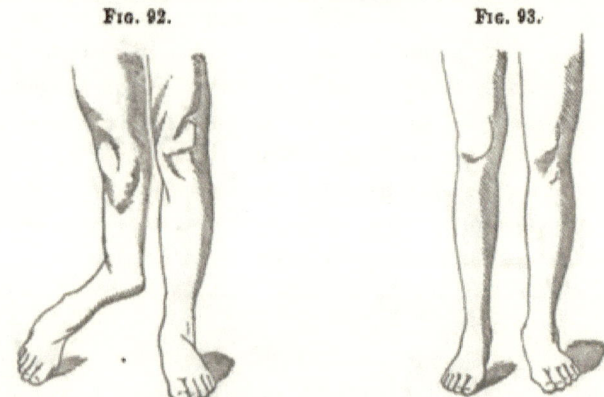

The above figures, engraved from photographs, represent the appearances of the limb: before treatment, Fig. 92; after treatment, Fig. 93.

It is certainly unfair to quote failures of this measure (of puncturing the soft callus between the fragments) as failures of the treatment by drilling. From the difficulty of puncturing the hard substance of the shaft of a long bone by an awl, it may be taken for granted that most of the cases in which such an instrument was improvised, were cases in which the callus alone was penetrated.*

* For much information with regard to the early methods of treatment, see an article by Dr. George W. Norris, in American Journal Medical Sciences, Oct. 1842, "On the Treatment of Deformities following unsuccessfully treated Fractures."

INDEX.

A.

Adams, Mr. **William, 97, 100, 102, 138, 175,** 189.

Althaus on Electricity, **40.**

Anæsthesia in **employing electricity, 42.**

    force applied under, 184.

Anal fissure, rationale of symptoms and treatment, **35.**

Apparatus for extension, 85.

    should it be applied immediately after division of tendons, 172.

Attachment for extension, Barwell's method, **94.**

Apparent dislocation, 83.

Abernethy, 60.

Acetabulum, necrosis, **55.**

    affection of, **in** hip disease, 81.

Adhesive plaster after burns, 76.

    applied, 89.

    for talipes, 182.

    to the leg, 196.

    to the foot, 199.

Andrews, Dr. Edmund, 55, 69, 91, 92, 113, 118, 121, 143.

Ankle, inflammation of, 85.

    weak, brace for, 206.

Accidents, 19–75.

Aching pain, 51.

Artificial tension, **the** complement of paralyzed muscles, 193.

Articulation, medio-tarsal, 165.

Anchylosis, 73.

    of hip joint, specimen in Dr. Pancoast's Museum, 82.

Alcoholic stimulants in spinal caries, 149.

Ambrose Paré, extension for vertical curvature, 148.

Antero-posterior curvature, 126.

Atmosphere, **150.**

Atrophy of muscles, 33, 100.

Angular curvature, 126.
Amputation, **spontaneous**, 25, 162.

B.

Bauer, Dr. Louis, 117, 143, 171, 181, 190.
Barwell, Mr. Richard, 28, 51, 52, 57, 63, 67, 80, 91, 92, 94, 96, 97, 168, 170,
        178, 180, 181, 182, 193, 198.
Bands, amniotic, 26.
Barton, 73.
Bladder, cases of deficient development of, 25.
        and **uterus,** palsying effect of over-distension, **37.**
Brace for a weak ankle, 206.
Brainard, Dr. Daniel, 74, 215, 219.
Bending bone, 214.
Beale, Dr. Lionel J., 62, 101–109.
Bed for extension, 112.
Bigg, Mr. H., apparatus for weak lower limbs, 44.
        apparatus for lateral curvature, 116.
Bonnet, M., 53, **54,** 71, 96, **119, 144, 168.**
Bones, softening **of, 76.**
        weakness of, in lateral **curvature of the spine, 98.**
Bow-legs, 209.
Boston Medical College **Museum, 82.**
Brodhurst, 112.
Brodie, Sir B. C., fractures unite in paralysis, 34, 62.
Bosch, M., 216.
Bouvier, M., 180.
Bouley, M., 180.
Brown-Sequard, Dr., 31, 33, 42, 185.
Burns, 75.
Bruises, 74.

C.

Calcaneus, **159.**
Calipers, **applied to** the foot, 201.
Caries of spine, apparatus, Dr. H. G. Davis, 141.
        support by lifting, 143.
Caries of vertebra, Pott's account, 129.
Cartilages destitute of sensibility, **52.**
Causes of talipes, 160.
Cautery, actual, 61.
Chair of Bonnet, 119.
        Andrews, 121.

Chest-counter-extension, 84, 94.
Chelius, 61, 62.
Classification, general, 19.
    of lateral curvatures of the spine, 98.
Cleft palate, 20.
Cicatrices cannot be elongated, 75, 211.
Cicatrix-contraction, 180.
Closure of the mouth, 76.
Cold and heat, for promoting capillary circulation, 43.
Coincidence, 164.
Compression of parts in utero, 162.
Copeland, Thomas, 128, 138, 144.
Complications of talipes, 160.
Contracture, 28, 57, 186.
Contraction, muscular, persistent, 28.
    voluntary and reflex, suspended by division and extension, 73.
Convulsions in infants, 168.
Correction of deformities after fractures, 214.
Couch, spinal, 113.
Counter-extension from the chest, 84–94.
Counter-irritation, 60.
Cruveilhier, illustration of angular curvature, 131.
Curvature of the spine, lateral, 96.
    posterior, 126.
    classification, 97.
    general remedies in, 134.
    local remedies, 135.
    counter-irritation in, 136.

D.

Davis, Dr. H. G., 64, 69, 85, 111, 118, 182.
Darwin's bed, 112.
Definition, 17.
Delay of union, 216.
De Le Barre, 21.
Delpech, apparatus for extension, 113.
Delpech, 172.
Detmold, Dr. W., 37, 74, 219.
Deviations at the knee-joint, 208.
Development, late in lower extremities, favorable to deformity, 164.
Deformities occasioned by perversion of nervous function, 27.
    from spasm and paralysis, more frequent in the lower extremities, 164.
    following cicatrices, 211.

Deformities in Pott's disease, explanation, 133.
  after fractures, 74, 213.
Diefenbach, 47, 215, 218.
Diet in caries of the spine, 148.
Disinfection, 150.
Dislocation, of the head of the femur, 81.
  apparent, 83.
Division, subcutaneous, in the vicinity of large arteries, not free from danger, 72.
  of muscles and tendons, explanation of the effects, 36.
Drilling bones, 74, 215.
Dorsals, 153.

E.

Earle, Sir James, 140.
Extension, exhausting and lengthening muscles, 37.
  before change of angle, 62, 69.
  not painful, 66.
  apparatus for, 85.
  attachment, Barwell's method, 94.
Elastic force, 210.
Elasticity useful in extension, 71.
Electric current, interrupted, for striped muscles, 30, 184. 186.
  continuous, excites smooth muscular fibre, 40.
Elongation, apparent, 80.
Equino-varus, 156.
Equinus, 151.
Esmarch on Cold, 135.

F.

Failure, owing to premature discontinuance of support, 38.
Fatigue of muscles in lateral curvature of the spine, 100.
Fractures unite in paralysis, 34.
Fracture of thigh followed by lesion of joints, 57.
Fragments of bone (minute), organize, 220.
Fergusson, 216.
Freezing, 74.
Friction in spinal curvature, 103.
Fingers, supernumerary, 26.
Finger-stalls, 210.
Fissura in ano, theory of it, and philosophy of cure, 35.
Fissures, congenital, 25.

Force under anæsthesia, 184.
Ford, 138.
Foster, J. Cooper, 63.

G.

Galvanism, 218.
Gnawing pain, 51.
Genu valgum, 208.
Gross, Dr. S. D., 171.
Guerin, 36.
Gutta-percha, applied to the foot, 191–200.

H.

Hamilton, case of, 223.
Hand, the type of apparatus for club-foot, 170, 206.
    a substitute for, a desideratum, 192.
Harris, Dr. Wm., 64.
Hart, Ernest, 48.
Heat and cold, for promoting capillary circulation, 42.
    sensitiveness to, in Pott's disease, 129.
Heel, in club-foot, to be depressed later, 202.
Hernia, principles of treatment, 24.
Hewins, Dr. L. T., experiments with divided tendons, 178.
Hip-disease, nervo-muscular phenomena in, 80.
Hippocrates, 187.
Hypertrophy as a cause of deformity, 77.
Hypospadias, 24.
Hossard, 115.

I.

Inactivity in spinal disease, 101.
Inflammation, 19, 49, 50.
    distortion and stiffening after, 67.
    of the knee-joint, 84.
Inheritance of peculiarities, 162.
Immediate union of bone, 221.
Interstitial fracture, 214.
Ischiatic crutch, 93, 94.
Immobility, line of, 212.
Iodide of iron, 135.

J.

Javal, M., 48.

K.

Key, Mr., 63.
Knee-joint, necrosis, 56.
    inflammation, 84.
    deviations at, 208.
Kingsley, N. W., 22.
Knock-knees, 208.
Kyphosis, 126.

L.

Lancing joints, 58.
Lateral curvature of the spine, 96.
    abrupt, specimen in cabinet of Medical College, Boston, 100.
Lateral deviation in fractures, 214.
Lateral curvature without a compensating curve, 125.
Leather employed by Hippocrates, 188.
    to the bottom of the foot, 208.
Leptandrin, 134.
Lewis, Dr. Dio, 103.
Ligaments, disease of, in lateral curvature of the spine, 99.
    susceptible of pain when stretched, 165.
Ling, movement cure, 103.
Line of immobility, 212.

M.

March, Dr. Alden, 63, 64.
Matteucci, experiment with galvanism, 36.
Mechanical treatment of paralyzed parts, 43.
    support in caries of spine in all stages, Adams, 139.
    treatment of talipes, 187.
Medicine and surgery must be studied together, 170.
Medio-tarsal articulation, 155, 165.
Mercury in Pott's disease, 134.
Metallic points, 220.
Michaelis, 172.
Mincius, 171.
Mitchel, Dr. S. Weir, 33.
Misplacements, 20, 26.

Mobility and extensibility, 167.
Morbus coxarius, 79.
    rarely leads to spontaneous dislocation, 82.
Motion, passive, to obviate atrophy, Brown-Sequard's theory, 34.
Mother, influence of mental impressions upon offspring, 162.
Movement-cure, 103.
Movements, localized, of Dr. C. F. Taylor, 39.
Mouth, closure of, 76.
Muscular balance unnecessary, 38.
    contraction, 53.
    curvatures of artificial production, 101.
Muscles, fatigue of, in lateral curvature of the spine, 100.
Muscle shortened as much as it can, will shorten no more, 186.
Mutilation, 19,·77.

N.

Nature, not to be imitated in the direction of force applied in club-foot, 194.
Nelaton, M., 215.
Nervo-muscular phenomena in hip-disease, 80.
Norris, Dr. G. W., 216.
Notta, 33.
Nutrition, perversions of, 49.
    in relation to symmetry, 161.
    deficient from obstruction of bloodvessels, ib.
    of contracted muscles, favored by division, 181.
    of muscles, favored by flexion and extension, ib.

O.

Obturation of De Le Barre, 21.
    of Stearne, ib.
Oesterlen, M., 216.
Overaction, prevention of, apparatus, 43.
Over-correction of strabismus, 46.
Over-extension, 73.
Offspring, influence of mental impressions of mother upon, 162.
Operations, plastic, 212.

P.

Paget's experiments in cutting tendons, 177.
Palate, cleft, 20.
    of vulcanized rubber, 22.

Pancoast, Dr. Joseph, 96.
    cabinet, 166.
Paralysis, with rigidity, 29, 30.
    explanation of it, 31.
    incomplete, 30.
    effects remain after removal of the cause, 32.
    with atrophy, 33.
    without rigidity, treatment, 39.
    as a therapeutic agent in spinal caries, 138.
    and spasm, relative importance of, in talipes, 165.
Palsy, Scribner's, 43.
    apparatus for self-exercise, 45.
Pain, 51.
Paré, Ambrose, 21, 78.
Parker, Dr. Willard, 54.
Plantaris, 157.
Plaster of Paris, 114.
Plastic operations, 212.
Perineal band, elastic, 86.
Periosteum, removal of, for establishing a line of immobility, 211.
Perversion, 19, 26.
Prevention, importance of, 17.
    of deformities after fractures, 213.
Pressure, a chief element of acute joint-inflammation, 65.
Prince's splint, 93.
    apparatus for lateral curvature, 123.
            for caries of spine, 146.
    plan for talipes, 203.
Prisms, plane, for strabismus, 48.
Physick, Dr. P. Syng, 64, 218.
Points, metallic, 220.
Posture in lateral curvature of the spine, 100.
Position of paralyzed parts, 43.
Post's gutta-percha shoe, 191.
Pott's disease, 126.
    medical treatment, 134.
    description of caries of vertebræ, 129.
Progressive shortening, 214.
Purgation in Pott's disease, 134.
Pus, evacuation of, 61.
    in spinal caries, 136.

Q.

Quietude in caries of the spine, 138.

R.

Redundancy, 20, 26.
Reid, Dr. John, experiment with electricity upon palsied muscles, 39, 184.
Removal of cicatrix, 75.
Resection, 218.
Rheumatism, 50.
Rickersteth, Mr., 216.
Rivet, for fracture, 220.
Rubber extension after burns, 76.
Rupture of tendo Achillis, 179.

S.

Sartorius, 172.
Sand, a bag of, carried on the head, in spinal curvature, 103.
Sayre, Dr. L. A., 73, 91.
Scarpa's Shoe, 189.
Shaw, Mr., 102.
Strabismus, how produced, 46.
    methods of operation, *ib.*
    treated by exercise of muscles, 47.
Staphyloraphy, success of, 20.
    reason of failure, 21.
Spasm, tonic, intra-uterine, 27.
    permanent, 33.
    in lateral curvature, 104.
    and paralysis produce deformities most frequently in the lower extremities, 164.
Separation of fragments, a cause of delay of union, 217.
Seton, 215, 218.
Screw for reducing deformity, 226.
Shelldrake, illustrations, 140.
Skey, Mr., 216.
Stearns, Dr. C. F., 21.
Stereoscope, for strabismus, 48.
Stretch-bed, 112.
Simpkins, case of, 226.
Skin, unnecessary pressure to be avoided, 206.
Spine, lateral curvature, 96.
Spinal shield of Bonnet, 144.
Splint, for extension, 86.
    long, 89.
    of Davis, 86

Splint, of Sayre, 91,
   of Barwell, *ib.*
   of Vedder, *ib.*
   of Andrews, 92.
   of Prince, 93.
   of Taylor, 92.
Stiffenings after inflammation, 67.
Spike, Malgaigne's, 221.
Syme, 174, 183.
Symmetry, impairment of, by accidental position, 161.
Synovial membranes, inflammation of, 49, 79.
Synovitis, spontaneous, 50.
     traumatic, *ib.*
     general treatment, 58.
     constitutional treatment, 59.
Strychnia, 187.
Soleus, contraction of, the cause of talipes, 166.
South, 61, 62.
Scoutetten's Shoe, 190.
Softening, of bone, 215.
     apparent, 80.
Shortening in fractures, 213.
     progressive, 214.
Shoe for extension, 95.
Skoliosis, 96.
Stromeyer, 37, 47, 171.
Support, artificial, to diseased vertebræ, 139.
Subcutaneous division of tendons by Stromeyer in 1831, 171.
Strumous synovitis, 50.

T.

Table of tenotomy, 176.
Talipes, definition, 151.
     classification, *ib.*
     complications, 160.
     causes, *ib.*
     treatment, 169.
     mechanical treatment, 187.
Tamplin, 71, 65.
Tamplin's case of division of tendons, 178.
Tartar emetic, 134.
Tavernier's belt, 116.
Taylor, Dr. C. F., 92, 121, 142.

Transverse diameter of the foot to be unfolded, 201

Tendo Achillis, **rupture,** subsequent union, 179.

Tendons, Paget's **experiments, 177.**

old rule for **division of, 173.**

Tenotomy, instruments for, 172.

Adams's table of, 176.

in **upper** extremity, 210.

Tension, artificial, the complement of **paralyzed** muscles, 193.

Tessier, M., 53, 54.

Thellenius, 172.

Tibia, dislocated, 68.

Tin, for adhesive plaster **attachment, 95.**

Twist of the spine, in lateral curvature, 97, 100.

Todd, Dr. **R. B., 29, 185, 186.**

Tonic **contraction of muscles, electricity of** doubtful application, 41.

Torticollis, **104.**

caused by irritation, 105.

U.

Ulceration of foot, 195.

Uterus and bladder paralyzed by over-distension, analogy of, 37.

Uterus, position of foot in, **a cause of talipes, 161.**

Union of parts of two or more **individuals, 162.**

of divided tendons, Paget's **experiments, 177.**

delay of, 216.

V.

Valgus, 156.

Varus, 154.

Van Buren, **Dr. Wm. H., 55.**

Vedder's **Splint, 91.**

Velpeau's expedient **for writers' palsy, 44.**

Vertical curvature, 126.

Vigilant repose, 102.

Volition, electricity a **substitute for, 185.**

Voluntary power as much **used to control, as to excite,** 168.

W.

Waist of the foot, 194.

Warren, Dr. J. **Mason, 20.**

Webster, Dr. Warren, case of rupture of tendo Achillis, and subsequent
        union, 179.
Weight upon the head in spinal curvature, 103.
Weinhold, 215.
Werner's anti-plastic movements, 111.
Wire breeches, 143.
Wiring bones, 215.
Wire for fractures, 220.
Wistar and Horner, Museum of, 82.
White swelling, 84.
Wrist joint, inflammation, 85.
Wry neck, 104, 107.
Wood, Dr., 115.
Wood and iron inefficient substitutes for the hand, 171.

                        Y.

Yielding force, 182.